Energy, Climate and the Environment

Series Editors
David Elliott
The Open University
Milton Keynes, UK

Geoffrey Wood
School of Law
University of Stirling
Stirling, UK

The aim of this series is to provide texts which lay out the technical, environmental and political issues relating to proposed policies for responding to climate change. The focus is not primarily on the science of climate change, or on the technological detail, although there will be accounts of this, to aid assessment of the viability of various options. However, the main focus is the policy conflicts over which strategy to pursue. The series adopts a critical approach and attempts to identify flaws in emerging policies, propositions and assertions. In particular, it seeks to illuminate counter-intuitive assessments, conclusions and new perspectives. The intention is not simply to map the debates, but to explore their structure, their underlying assumptions and their limitations. The books in this series are incisive and authoritative sources of critical analysis and commentary, clearly indicating the divergent views that have emerged whilst also identifying the shortcomings of such views. The series does not simply provide an overview, but also offers policy prescriptions.

More information about this series at
http://www.palgrave.com/gp/series/14966

Victoria R. Nalule

Energy Poverty and Access Challenges in Sub-Saharan Africa

The role of regionalism

palgrave
macmillan

Victoria R. Nalule
Centre for Energy, Petroleum and Mineral Law
and Policy (CEPMLP)
University of Dundee
Scotland, UK

Energy, Climate and the Environment
ISBN 978-3-030-07029-8 ISBN 978-3-319-95402-8 (eBook)
https://doi.org/10.1007/978-3-319-95402-8

Cover illustration: Petri Oeschger/gettyimages

This Palgrave Macmillan imprint is published by the registered company Springer Nature Switzerland AG
The registered company address is: Gewerbestrasse 11, 6330 Cham, Switzerland

Preface

Energy access has gained a lot of attention following the adoption of the Sustainable Development Goals (SDG), which under goal 7 emphasises the access to affordable, reliable, sustainable, and modern energy for all. Energy is central in addressing the major global challenges of the twenty-first century, including poverty, climate change, and famine to mention but a few; and yet a large number of people especially in Sub-Saharan Africa (SSA) still lack access to modern energy, such as electricity, natural gas, and liquefied natural gas (LNG) for lighting and cooking, respectively. Initiatives have been taken at an international, regional, and national level to address the issue of energy access; however, the problem continues to escalate in SSA. SSA countries have common energy challenges which could be tackled with common efforts through regional cooperation. Achieving cooperative outcomes can be difficult among countries as such at a bilateral level. In this regard, the employment of a broader multi-country framework, which can be utilised through the existing regional organisations and international organisations, might be expected to lead to greater cooperation in the energy sector. This possibility is the springboard for this book.

This book highlights the key regional energy cooperative mechanisms employed by SSA regional organisations and demonstrates if these are efficient to tackle the challenge of energy access.

Scotland, UK — Victoria R. Nalule

Acknowledgements

First of all, I return all glory, honour, and adoration to the almighty God. I am deeply indebted to Professor Peter Cameron who provided comments on my PhD thesis, which is the basis for this book. I also thank Dr Ernesto Bonafè for his comments.

My sincere gratitude goes to my Dad Migadde Robert Blick and the entire family for supporting me. I would like to thank Dr Geoffrey Woods for his proofreading and comments on this work.

I would like to thank the staff at the Southern African Development Community Secretariat (SADC), especially Remious Makumbe, Moses Ntlamelle, and Odala Matupa, for their assistance and support while I was carrying out research at SADC. I would also like to thank the staff at the East African Community Secretariat (EAC) especially Deputy Secretary General Hon. Christopher Bazivamo, Lawrence Mukwaya, and Elsam Byempaka for all their support.

I also sincerely thank the staff at the Energy Charter Secretariat who played a big role in shaping my career, specifically Secretary General Urban Rusnák and Dr Ernesto Bonafè, who were supportive during my stay in Brussels.

To all my friends, thank you for being such magnificent people. I love and cherish you all!

Contents

Abbreviations

ACER	The Agency for the Cooperation of Energy Regulators
ACPEU	The African Caribbean Pacific-European Union Partnership Agreement
AELG	African Energy Leaders Group
AfDB	African Development Bank
AGECC	Advisory Group on Energy and Climate Change
AUC	African Union Commission
CEER	Council of European Energy Regulators
DoE	Department of Energy
DRC	Democratic Republic of Congo
EAC	East African Community
ECOWAS	Economic Community of West African States
ECT	Energy Charter Treaty
EnDev	Energising Development
ERA	Energy Regulatory Authority
ERB	Energy Regulation Board of Zambia
ERERA	ECOWAS Regional Electricity Regulatory Authority
ERF	Electricity Regulatory Forum
ERGEG	European Regulatory Group for the Electricity and Gas
ESMAP	Energy Sector Management Assistance Program
ESWG	Energy Sustainability Working Group
EWURA	Energy and Water Utilities Regulatory Authority of Tanzania
FERC	Federal Energy Regulatory Commission

GDP	Gross Domestic Product
GRF	Gas Regulatory Forum
GTF	Global Tracking Framework
ICS	Improved biomass cook stove
IEA	International Energy Agency
IEC	International Energy Charter
IRSE	Institute for Electricity Sector Regulation of Angola
JAEI	Japan-Africa Energy Initiative
JPOI	Johannesburg Plan of Implementation
LEWA	Lesotho Electricity and Water Authority
LNG	Liquefied natural gas
LPG	Liquefied petroleum gas
MDGs	Millennium Development Goals
MERA	Malawi Energy Regulatory Authority
NER	National Electricity Regulator
NERSA	National Energy Regulator of South Africa
NIC	Namibia Investment Center
NPCA	NEPAD Planning and Coordinating Agency
NRAs	National Regulatory Authorities
OOC	Oversight and Operations Committee
PUC	Public Utilities Corporation
REC	Regional Economic Communities
RERA	Regional Electricity Regulatory Authority
RISE	Readiness for Investment in Sustainable Energy
SADC	Southern Africa Development Community
SDGs	Sustainable Development Goals
SEC	Swaziland Electricity Company
SE4All	Sustainable Energy for All
SERA	Swaziland Energy Regulatory Authority
SIB	Seychelles Investment Board
SSA	Sub-Saharan Africa
TISA	Trade and Investment South Africa
UN	United Nations
UNDP	United Nations Development Program
WHO	World Health Organization
ZERA	Zimbabwe Energy Regulatory Authority

List of Figures

List of Tables

1

Introduction

1.1 Introduction to Energy Access and Regionalism

It has long been recognised that energy is central to addressing the major global challenges of the twenty-first century including poverty, climate change, famine, and environmental degradation, to mention but a few. This is due to the fact that energy is not only essential for the provision of basic social services such as education and health care services, but it is also essential for industrialisation and the general economic development. Africa is a curious example of a continent (although by no means the only example) that is home to vast energy resources yet today is also home to hundreds of millions of people still lacking even basic access to modern energy such as electricity, natural gas, and liquefied natural gas (LNG) for lighting and cooking, respectively.

Modern energy can be distinguished from traditional energy by looking at the quality of energy used for instance with regard to traditional energy candles, kerosene, and lamps are used for lighting; and firewood for cooking. Over-reliance on traditional energy and having limited/no access to modern energy is basically what is referred to as "energy poverty"

V. R. Nalule, *Energy Poverty and Access Challenges in Sub-Saharan Africa*, Energy, Climate and the Environment, https://doi.org/10.1007/978-3-319-95402-8_1

and in some instances energy access. In reality the concept of energy access and energy poverty is complex, especially with regard to understanding or agreeing on where energy poverty ends or what energy access really entails and means. Issues may arise as to whether a household which shifts from firewood and adopts modern stoves for cooking and a small solar panel for lighting up a few bulbs can qualify as one with access to modern energy: And if so, what about energy for business such as that needed in industries and large-scale farming? These issues highlight the universal difficulty in addressing energy access challenges, thus the increasing number of people lacking access to modern energy in Sub-Saharan Africa (SSA).

This is despite numerous initiatives taken over decades at the international, regional, and national levels to address the problem of energy access. Of relevance to this book, the problem not only persists but continues to escalate in SSA, a major region of Africa covering approximately 27,000 square kilometres. Geographically, SSA consists of the part of the African continent that lies south of the Sahara Desert. Politically, it consists of all African countries that are fully or partially located south of the Sahara, with the notable exception of Sudan. In energy terms, the SSA region is home to abundant energy resources including both fossil fuels and renewables. Despite the energy "wealth" of these countries, most of their citizens remain poor with the majority having no access to electricity.

It is against this stark background that this book looks closely at the regional efforts taken in SSA to tackle the challenge of energy access and energy poverty. There is no doubt that SSA countries have common energy challenges including inadequate and unreliable electricity supply, and heavy reliance on inefficient forms of energy such as traditional biomass—which has various negative effects on health and environment. For instance in the Southern African Development Community (SADC) region, it has been reported that more than 153,000 people die each year from household pollution resulting from indoor burning of solid fuels such as traditional biomass for cooking and heating.[1] These common energy challenges could be tackled with common efforts through regional

[1] See SADC Energy Monitor, 2016, Page 58.

cooperation. Achieving cooperative outcomes, however, can be difficult in practice. As such, the existence of a broader framework, which can be achieved through existing regional and international organisations, can lead to greater cooperation in the energy sector.

The focus of this book therefore is on the regional cooperative mechanisms employed by SSA regional organisations. There are a number of existing cooperative mechanisms already in place in SSA regional organisations, including the East African Community (EAC), the Economic Community of West African Countries (ECOWAS), and the Southern African Development Community (SADC). These regional organisations are the main case studies of this book and these are discussed in the next section:

1.1.1 The East African Community (EAC)

The EAC is a regional inter-governmental organisation of six countries.[2] It was established under Article 2 of the EAC Treaty with the main objective of widening and deepening cooperation among partner states. The EAC regional integration is a progressive process through four major pillars: Customs Union, Common Market, Monetary Union, and the Political Federation. Moving from one level of integration to another is through negotiated Protocols. In 2004, a Protocol for the establishment of the East African Custom Union was signed by the three original partner states and entered into force in 2005. Rwanda and Burundi joined the Custom Union in 2009. This was followed by establishment of a common market, which entered into force in 2010. The Protocol for establishing a Monetary Union was signed in November 2013 and its operationalisation is ongoing with a ten-year implementation phase before adopting a single currency for the region. The EAC has five organs and these are responsible for ensuring effective operation of the EAC. These organs include: the Summit comprising of heads of government of partner states, the Council of Ministers, coordinating Committee,

[2] EAC is a regional inter-governmental organisation of six countries including Burundi, Rwanda, Kenya, Tanzania, South Sudan, and Uganda.

Sectoral Committees, the East African Court of Justice, the East African Legislative Assembly, and the Secretariat.[3]

1.1.2 The Economic Community of West African Countries (ECOWAS)

ECOWAS is a regional inter-governmental organisation of 15 countries.[4] It was established under Article 2 of the ECOWAS Treaty, with the main objective of promoting cooperation and integration in West Africa. Though ECOWAS was established in 1976, the desire to create regional cooperation among West African states dates back to 1964, with a call from President William Tubman of Liberia. This followed the signing of an Agreement among Cote d'Ivoire, Guinea, Liberia, and Sierra Leone in 1965 and later the adoption of the Protocol establishing ECOWAS on 5 November 1976 by the 15 member states. ECOWAS consists of the Authority of Heads of State and Government, the Council of Ministers, the Community Tribunal, Community Court of Justice, the Executive Secretariat, the ECOWAS Parliament, and the Specialised Commissions.

1.1.3 The Southern African Development Community (SADC)

SADC is a regional inter-governmental organisation of 15 countries.[5] It was established under Article 2 of the 1992 Treaty with the main objective of achieving development, economic growth, and alleviation of poverty through regional integration.[6] SADC is the successor to the Southern African Coordinating Conference (SADCC). SADCC was formed in

[3] The Organs and Institutions of the Community are established under Article 9 of the EAC Treaty.

[4] ECOWAS is a regional inter-governmental organisation of fifteen countries including Benin, Burkina Faso, Cabo Verde, Côte d'Ivoire, the Gambia, Ghana, Guinea, Guinea Bissau, Liberia, Mali, Niger, Nigeria, Senegal, Sierra Leone, and Togo.

[5] SADC is a regional inter-governmental organisation of fifteen countries including Angola, Botswana, Democratic Republic of Congo, Lesotho, Madagascar, Malawi, Mauritius, Mozambique, Namibia, Seychelles, South Africa, Swaziland, United Republic of Tanzania, Zambia, and Zimbabwe.

[6] Article 5(1) (a) of the 1992 SADC Treaty.

Lusaka, Zambia in 1980 with the adoption of the Lusaka Declaration.[7] SADCC was formalised by a Memorandum of Understanding (MOU) on the Institutions of the Southern African Development Coordination Conference dated July 1981. In an effort to formalise SADCC and give it an appropriate legal status, the MOU was replaced by a Treaty. In this regard, the SADC Declaration and Treaty was signed on 17 August 1992 and this effectively transformed SADCC into SADC.[8] Six institutions are established under the SADC Treaty. These include the Summit of Heads of States or Governments, the Council of Ministers, Commissions, the standing Committee of officials, the Secretariat, and the Tribunal. The Summit is the supreme policy making institution of SADC; it consists of the Heads of States or Governments of each member state, and it is responsible for the overall policy direction and control of the functions of SADC.[9]

Looking at EAC, ECOWAS, and SADC provides a unique insight in the cooperative activities in various regions of SSA including the East, West, and South. These regions are also home to massive energy resources including both renewables and fossil fuels. The unique feature with regard to the achievements, number of member states, and organisation of these regional organisations presents a good ground to carry out a comparative analysis of their regional efforts in tackling the challenge of energy poverty. Moreover, their historical background in terms of the reasons for their establishment, for instance the political reasons for SADC and the economic reasons for both ECOWAS and EAC, also presents a good insight on the aspect of regionalism in SSA.

In particular, the key issue addressed in this book is whether the energy regional cooperative mechanisms employed by EAC, ECOWAS, and SADC are effective in tackling the challenge of energy access in SSA. The book basically makes a unique contribution by providing a contemporary, comprehensive scholarly and practical assessment of energy access in SSA through the lens of these regional organisations. In addressing the

[7] SADCC was formed on April 1 1980, comprising of all the majority ruled states of Southern Africa: Angola, Botswana, Lesotho, Malawi, Mozambique, Swaziland, Zambia, and Zimbabwe.

[8] The Summit was held in Windhoek, Namibia, the SADC Treaty was signed by the Heads of State and Government, and this established SADC under Article 2.

[9] Institutions are established under Article 9(1) of the Treaty of the Southern African Development Community.

issue of energy access, a legal and comparative analysis of the regional energy cooperative mechanisms including regional regulatory authorities, regional energy infrastructure establishments, and regional renewable energy initiatives and developments is undertaken.

1.2 Why Solving Energy Poverty Is Essential in SSA

The concentration of expertise and funding in SSA regional organisations and their ready access to them suggests that they can spearhead the fight against energy poverty by encouraging more and effective regional cooperation in the energy sector. This can however be achieved only if there is effective regional cooperation in the energy sector of various African countries.

To clearly appreciate the problem of energy access in SSA, we need to understand that energy is essential in achieving most of the Sustainable Developmental Goals (SDGs). Basically, the SDGs are a new set of universal goals which will guide United Nations (UN) member states in their agendas and political policies. The SDGs are 17 in number and they follow and expand on the Millennium Development Goals (MDGs). Although the MDGs were global goals, they seemed to mostly focus on developing countries (in as far as the relevance of goals was concerned). Paradoxically, some reports indicate that most of the poor countries in SSA did not meet the MDGs in 2015. Some of the reasons which have been highlighted for the failure of MDGs are that these goals were too narrow and did not consider the root causes of poverty. Moreover, other scholars have highlighted how the measurement of "success" or "failure" in meeting the MDGs was unfair to some developing countries.[10] In this section, the focus will be on the SDGs in as far as bringing out the argument that access to modern energy is essential in meeting most of the SDGs, and as such it can be asserted and argued that, as long as the

[10] For a full discussion on this see Easterly, W., 2009. How the millennium development goals are unfair to Africa. *World Development, 37*(1), pp. 26–35; See also Attaran, A., 2005. An immeasurable crisis? A criticism of the Millennium Development Goals and why they cannot be measured. *PLoS Medicine, 2*(10), p. 318.

challenge of energy access persists, then many other global challenges will be hard to solve in SSA.

1.2.1 Link Between Energy Access and Other SDGs

Though the SDGs are highly recognised as capable of addressing the pressing global problems, these have also been harshly described by Lancet, who referred to them as "*fairy tales, dressed in the bureaucratese of intergovernmental narcissism, adorned with the robes of multilateral paralysis, and poisoned by the acid of nation-state failure.*"[11] Lancet argues that there is a need to measure the full meaning of sustainability if the SDGs are to be successful.[12] Additionally, other scholars argue that there is a need to pay attention to the inter-linkages in the three areas including across sectors[13]; across societal actors[14]; and between and among low-, medium-, and high-income countries.[15] In this section, a link between energy access and the other SDGs will be discussed. Prior to the discussion, it is worth noting that energy access is also among the SDGs and this is provided for under goal 7, which advocates for access to affordable, reliable, sustainable, and modern energy for all.[16]

1.2.1.1 Energy Access and Poverty Eradication

The first of the SDGs is to "*end poverty in all its forms everywhere*". Basically, energy access is a major problem because it has escalated the level of poverty in the SSA region. It is however important to note that

---

[11] See Horton, R., 2014. Offline: Why the sustainable development goals will fail. *The Lancet, 383*(9936), p. 2196.

[12] Horton, R., 2014. Offline: Why the sustainable development goals will fail. *The Lancet, 383*(9936), p. 2196.

[13] Sectors such as finance, agriculture, energy, and transport.

[14] Societal actors such as local authorities, government agencies, private sector, and civil society.

[15] For a full discussion on this, see Stafford-Smith, M., Griggs, D., Gaffney, O., Ullah, F., Reyers, B., Kanie, N., Stigson, B., Shrivastava, P., Leach, M. and O'Connell, D., 2017. Integration: The key to implementing the Sustainable Development Goals. *Sustainability Science, 12*(6), pp. 911–919.

[16] Goal 7 of the UN Sustainable Development Goals, 2016.

lack of access to modern energy may not be considered by other scholars as the main cause of poverty in developing countries. There are other factors that ought to be considered including corruption, poor governance, lack of economic diversification, inequality in resource distribution, and many other factors some of which may be unique to a particular country in question, for instance political instability.

It has also been recognised that improved forms of energy are not sufficient conditions for development and as such other factors including infrastructure developments need to be considered.[17] There are also arguments that poverty just like apartheid and slavery is man-made and as such human solutions can help in eradicating it completely in Africa.[18] Additionally, other scholars have highlighted gender inequality in education as a source of poverty and low economic growth in developing countries.[19]

Although the number of people living on less than USD 1.90 a day fell from 57% in 1990 to 43% in 2012, the number of poor people on the other hand increased by more than 100 million (from 288 to 389 million). Moreover, the Gross Domestic Product (GDP) is also estimated to have declined in 2015, to 3.7%, from 4.6% in 2014.[20] This therefore proves that poverty is escalating globally, thus making it a major issue to be addressed nationally, regionally, and internationally.

Generally, in this book, the argument is that the global goal of eradicating poverty cannot be achieved without tackling the challenge of energy access, and this is due to the fact that there is a strong link between the two as recognised by the World Summit on Sustainable Development.[21]

---

[17] Cecelski, E. and Policy, A.A.E., 2002. Enabling equitable access to rural electrification: Current thinking and major activities in energy, poverty and gender. *World Development Report, 1*, pp. 2–3.

[18] See Nelson Mandela's speech in London's Trafalgar Square for the campaign to end poverty in the developing world, 2005. It can be accessed at http://news.bbc.co.uk/1/hi/uk_politics/4232603.stm

[19] See Abu-Ghaida, D. and Klasen, S., 2004. The costs of missing the Millennium Development Goal on gender equity. *World Development, 32*(7), pp. 1075–1107; Ansell, N., 2017. Achieving gender parity in education: Achievements and limitations of Millennium Development Goal 3. *Laboring and Learning*, pp. 111–130.

[20] World Bank: Overview, http://www.worldbank.org/en/region/afr/overview#1, last visited on 21 November 2015.

[21] UN-Energy: *The energy challenge of achieving the millennium development goals*, http://www.un-energy.org/publications/50-the-energy-challenge-for-achieving-the-millennium-development-goals, last visited on 11 November 2015.

However, besides modern energy we have to also consider other essential factors such as ensuring gender equality that have been highlighted by some scholars as being key in eradicating poverty in Africa: Gender has been linked to poverty to the extent that poverty has been feminised by some scholars. The characteristics of the feminisation of poverty have been stated to include among others that women experience a higher incidence of poverty than men; women face more barriers to lifting themselves out of poverty; women experience greater depth of poverty than men; and that women headed households are the "poorest of poor".[22]

Technology has also been highlighted as being essential in eradicating poverty. For instance, in his book *The end of poverty: How we can make it happen in our lifetime* Professor Jeffrey D. Sachs argues that technology is key in eradicating poverty. Referring to Japan's success story, he states that "*I believe that the single most important reason why prosperity spread, and why it continues to spread, is the transmission of technologies and the ideas underlying them. Even more important than having specific resources in the ground, such as coal, was the ability to use modern science-based ideas to organize production*".[23]

All these factors considered by various scholars including Professor Jeffrey Sachs are indeed essential in eradicating poverty, but one has to ask a few questions: How can technology function without basic energy such as electricity? How can we achieve industrialisation, if the industries lack modern energy? In this regard therefore we can ably argue that access to modern energy is key in eradicating poverty in SSA and other developing regions and thus essential in achieving goal one of the SDGs.

1.2.1.2 Energy Access and Hunger Eradication

The second SDG is to end hunger, achieve food security and improved nutrition, and promote sustainable agriculture. The link between energy and agriculture is the ability of small-scale farmers in developing countries

[22] For a full discussion on this see Chant, S.H., 2007. *Gender, generation and poverty: Exploring the feminization of poverty in Africa, Asia and Latin America*. Edward Elgar Publishing.

[23] For a full discussion on this see Sachs, J., 2005. *The end of poverty: How we can make it happen in our lifetime*. Penguin UK.

to use electricity and diesel for irrigation. In India for instance, there are about 84.3% small farm holders who require secure water supply through irrigation to grow their crops.[24] Often, small-scale farmers in SSA just like in India use ground water for irrigation—by using pumps which are operated by diesel or electricity.[25] Rahman and Bhatt while advocating for the use of solar energy in the Indian agriculture sector argue that "*the energy squeeze in the country undermines the adoption of broad range precision irrigation technologies and stable and profitable productivity of small scale farmers*".[26] The lack of access to energy therefore implies that many small-scale farmers and their families will be at the risk of dying of hunger and the agricultural sector will also be heavily affected.

Additionally, besides irrigation, some farmers also use agricultural tools and machines which require energy such as electricity or diesel to operate. In this regard, there is enormous literature which argues that electricity as a form of energy is essential in the irrigated agriculture sector, largely due to the fact that this sector has become more modernised and as such the rising energy costs affect irrigated farming.[27] Moreover, the connection between energy and water is also well documented.[28] Electricity is also key in the food preservative process for instance through refrigeration. This therefore makes access to modern energy essential in achieving the UN goal two of ending hunger.

1.2.1.3 Energy Access and Health

Ensuring healthy lives is the third goal of the SDGs. There are various health challenges including among others child deaths, maternal deaths,

[24] Rahman, A. and Bhatt, B.P., 2017. Scope and options of solar energy use in agriculture in eastern region of India. *Journal of AgriSearch, 4*(1).

[25] Ibid.

[26] Ibid.

[27] Langarita, R., Chóliz, J.S., Sarasa, C., Duarte, R. and Jiménez, S., 2017. Electricity costs in irrigated agriculture: A case study for an irrigation scheme in Spain. *Renewable and Sustainable Energy Reviews, 68*, pp. 1008–1019.

[28] Lecina, S., Isidoro, D., Playán, E. and Aragüés, R., 2010. Irrigation modernization and water conservation in Spain: The case of Riegos del Alto Aragón. *Agricultural Water Management, 97*(10), pp. 1663–1675; Rodríguez-Díaz, J.A., Pérez-Urrestarazu, L., Camacho-Poyato, E. and Montesinos, P., 2011. The paradox of irrigation scheme modernization: More efficient water use linked to higher energy demand. *Spanish Journal of Agricultural Research, 9*(4), pp. 1000–1008.

and epidemic diseases such as AIDS, malaria, and tuberculosis.[29] Estimates indicate that more than six million children die before their fifth birthday each year.[30] In this regard, one of the UN target goals is to end preventable deaths of newborns and children under five years of age.[31]

The connection between energy and healthy living lies in the fact that lack of access to modern energy implies that many people in SSA will use traditional energy such as firewood for cooking and this definitely has various negative effects to the lives of the people especially due to the indoor air pollution. The IEA report indicates that household air pollution due to over-reliance on traditional energy is responsible for 2.8 million premature deaths every year.[32] In fact, one of the targets for goal three is "to reduce the number of deaths and illnesses from hazardous chemicals and air, water, and soil pollution and contamination".[33] Additionally, health care services such as hospitals require energy such as electricity to operate effectively, and as such it becomes hard to achieve the third goal of SDGs if the majority of people have no access to modern energy.

1.2.1.4 Energy Access and Quality Education

Goal 4 of the SDGs is quality education. Having and obtaining quality education is considered to be essential in improving people's lives and sustainable development. However, estimates indicate that 57 million children remain out of school in developing countries and more than half of them are from SSA.[34] Modern energy such as electricity has the potential of ensuring the achievement of quality education because this plays a

---

[29] See UN Sustainable Development Goals at http://www.un.org/sustainabledevelopment/health/, last visited on 12 January 2018.

[30] Ibid.

[31] See UN Sustainable Development Goals at http://www.un.org/sustainabledevelopment/health/, last visited on 12 January 2018.

[32] See International Energy Agency: World Outlook Special Report, 2017, Page 28.

[33] See UN Sustainable Development Goals at http://www.un.org/sustainabledevelopment/health/, last visited on 12 January 2018.

[34] See UN Sustainable Development Goals at http://www.un.org/sustainabledevelopment/education/, last visited on 12 January 2018.

big role in ensuring that facilities such as laboratories, libraries, and computers are adequately operating. An attempt to increase the quality of education through technology was the programme of "One Laptop per Child" (OLPC), which aimed at providing laptops to millions of children around the world including those in poor countries. Considering the social and economic challenges in poor countries such as lack of access to electricity, the OLPC programme was criticised, and to this effect, one scholar was of the view that "*provision of individual laptops is a utopian vision for the children in the poorest countries, education and social futures could be more effectively improved if the same investments were instead made on more sustainable and proven interventions.*"[35] Moreover, students would also require efficient light to carry out their revisions. This therefore proves that access to modern energy plays a big role in ensuring access to quality education in SSA, which is goal 5 of the SDGs.

1.2.1.5 Energy Access and Gender Equality

Goal 5 of the SDGs is on gender equality. The link between energy and gender equality is well documented especially considering the fact that women are the major users of energy resources in developing countries. It has thus been submitted that women in developing regions such as SSA experience energy challenges differently and more severely from their male counterparts.[36] Given the social roles of women in various SSA countries, such as cooking and other domestic work, women spend a lot of time collecting biomass fuels such as firewood: Estimates indicate that women spend on average 1.4 hours a day collecting fuel wood and four hours for cooking.[37]

This is indeed one of the reasons why some scholars have argued that access to energy is gendered.[38] Ensuring access to modern energy therefore

[35] See Warschauer, M. and Ames, M., 2010. Can one laptop per child save the world's poor? *Journal of International Affairs, 64*(1), pp. 33–51.

[36] Danielsen, K., 2012. *Gender equality, women's rights and access to energy services.* A paper presented to the Ministry of Foreign Affairs of Denmark.

[37] See International Energy Agency: World Energy Outlook Special Report, 2017, Page 3. The report can be accessed at https://www.iea.org/publications/freepublications/publication/WEO2017SpecialReport_EnergyAccessOutlook.pdf, last visited on 1 March 2018.

[38] For a full discussion on this see Danielsen, K., 2012. *Gender equality, women's rights and access to energy services.* A paper presented to the Ministry of Foreign Affairs of Denmark.

has the potential of improving the livelihoods of several women who depend on these energy resources for their day-to-day activities in developing countries, and as such contributes to achieving the SDG target of gender equality.

1.2.1.6 Energy Access and Economic Growth: With a Focus on the Mining Sector

Goal 8 of the SDGs emphasises the need to promote economic development and decent work for all. In this regard, one of the targets to be met by 2030, besides productive decent work for all, is the formalisation of micro-, small-, and medium-sized enterprises.[39] With respect to small-scale enterprises, the focus of this section will be on artisanal mining in SSA and how access to modern energy by artisanal miners can contribute in achieving goal 8 of the SDGs. I had a chance of carrying out field work research in the Ugandan and Tanzanian artisanal mines and my conclusion from this fieldwork is that lack of advanced technology (which is also limited due to lack of access to modern energy) has forced many artisanal miners to work in life threatening conditions; and these conditions will make it hard for the full achievement of goal 8 of the SDGs.

Energy Access and Artisanal Gold Mining

In artisanal gold mining (AGM) many miners still use rudimentary methods such as amalgamation to extract gold.[40] Gold amalgamation necessitates the use of mercury, which has various health and environmental impacts. Actually, it is estimated that 1400 tons of mercury is released into the environment from AGM.[41]

---

[39] See target 8.3 of the SDGs, https://sustainabledevelopment.un.org/topics/sustainabledevelopmentgoals last accessed on 14 January 2018.

[40] Gold amalgamation is one of the oldest forms of extracting gold: mercury is added to silver and gold to form an amalgam (paste) and it amalgamates all metals except iron and platinum. For a full discussion on this see Veiga, M.M., Angeloci-Santos, G. and Meech, J.A., 2014. Review of barriers to reduce mercury use in artisanal gold mining. *The Extractive Industries and Society*, *1*(2), pp. 351–361.

[41] See Assessment, U.G.M., 2013. *Sources, emissions, releases and environmental transport* (p. 42). Geneva, Switzerland: UNEP Chemicals Branch. Effects of mercury are well documented and for

Even though there are various issues associated with AGM, we cannot ignore the fact that it has the potential of improving the economic growth of many developing countries and this is due to the fact that AGM employs a big number of local people. As of 2012, AGM accounted for approximately 12% of all the gold produced in the world.[42] However, there is an urgent need to improve the working conditions of the workers involved in AGM, and one of the ways is to ensure access to advanced technology such as machinery, which will definitely require modern energy to fully be operative.[43] Of course, other ways of improving the lives of AGM workers have been discussed including suggestions that mercury pollution in AGM can be controlled through formalisation of the artisanal miners.[44]

Energy Access and Artisanal Salt Mining

Besides AGM, the other extractive sector which is capable of ensuring the achievement of goal 8 of the SDGs is artisanal salt mining. In Uganda for instance, there are several men and women involved in the artisanal mining of salt from Lake Katwe, and this activity has been

further discussions see also Pfeiffer, W.C., de Lacerda, L.D., Malm, O., Souza, C.M.M., da Silveira, E.G. and Bastos, W.R., 1989. Mercury concentrations in inland waters of gold-mining areas in Rondonia, Brazil. *Science of the Total Environment, 87*, pp. 233–240; van Straaten, P., 2000. Human exposure to mercury due to small-scale gold mining in northern Tanzania. *Science of the Total Environment, 259*(1), pp. 45–53; Spiegel, S.J. and Veiga, M.M., 2010. International guidelines on mercury management in small-scale gold mining. *Journal of Cleaner Production, 18*(4), pp. 375–385.

[42] Veiga, M.M., Angeloci-Santos, G. and Meech, J.A., 2014. Review of barriers to reduce mercury use in artisanal gold mining. *The Extractive Industries and Society, 1*(2), pp. 351–361.

[43] To have a general view of the living conditions of artisanal gold miners, take a look at some of the videos I shot from the field; this can be accessed on Victoria Nalule YouTube Channel, https://www.youtube.com/watch?v=HuWmudeSz_A&t=32s, last visited on 13 January 2018.

[44] See Veiga, M.M., Angeloci-Santos, G. and Meech, J.A., 2014. Review of barriers to reduce mercury use in artisanal gold mining. *The Extractive Industries and Society, 1*(2), pp. 351–361. Other scholars have also discussed the issue of formalisation of AGM; see Hinton, J.J., Veiga, M.M. and Veiga, A.T.C., 2003. Clean artisanal gold mining: A utopian approach? *Journal of Cleaner Production, 11*(2), pp. 99–115; McDaniels, J., Chouinard, R. and Veiga, M.M., 2010. Appraising the Global Mercury Project: An adaptive management approach to combating mercury pollution in small-scale gold mining. *International Journal of Environment and Pollution, 41*(3–4), pp. 242–258.

traditionally practised for decades. However, the miners don't have a decent working environment and as such they work in life threatening conditions such as being exposed to the ammonium gas which escapes from the salt water thus irritating their private parts.[45] In an endeavour to ensure a decent working condition for salt miners, some scholars have suggested employment of advanced technology by these artisanal salt miners, including the use of solar evaporation technique and mechanisation.[46]

Energy Access and Sand Mining

Another extractive industry activity which has become a major topic in recent years is sand mining, and this definitely not only has a role in ensuring the achievement of goal 8 regarding economic growth, but it is also essential in the achievement of goals 9 and 11 of SDGs relating to infrastructure and sustainable cities and communities, respectively.[47] Sand is one of the most mined and demanded material in the twenty-first century and this is due to the fact that it is a main ingredient of concrete, plaster, and asphalt paving, which are used in the construction industry.[48] Goal 11 of the SDGs advocates for sustainable cities and communities; the construction of these cities and various infrastructure will require enormous amounts of sand. Moreover, it is estimated that more than half of the world's population now live in cities:

---

[45] To get a full idea of the challenges artisanal salt miners face in Uganda, watch a field work video I recorded. This can be accessed at Victoria Nalule YouTube Channel, Salt Mining in Uganda, https://www.youtube.com/watch?v=RKFP6DswMrM&t=532s, last visited on 14 January 2018. In this video, there are interviews conducted with the artisanal salt miners themselves. The video also highlights the working conditions of the miners.

[46] See Kirabira, J.B., Kasedde, H. and Semukuuttu, D., 2013. Towards the improvement of salt extraction at Lake Katwe. *International Journal of Scientific & Technology Research, 2*(1), pp. 76–81. See also, Sampathkumar, K., Arjunan, T.V., Pitchandi, P. and Senthilkumar, P., 2010. Active solar distillation—A detailed review. *Renewable and Sustainable Energy Reviews, 14*(6), pp. 1503–1526.

[47] See UN Sustainable Development Goals, https://sustainabledevelopment.un.org/topics/sustainabledevelopmentgoals, last accessed on 14 January 2018.

[48] See Aljazeera, Environment, "The World is running out of sand", http://www.aljazeera.com/video/news/2017/06/world-running-sand-scientists-170625060441463.html, last visited on 14 January 2018; The New Yorker, https://www.newyorker.com/magazine/2017/05/29/the-world-is-running-out-of-sand, last visited on 14 January 2018.

and that China has used more concrete in the past few years than the United States used in the entire twentieth century.[49]

Despite reports indicating that the world is running out of sand, not much has been done in SSA countries to preserve this resource, and paradoxically some people who stay in areas endowed with sand are not aware of the importance of this resource. Most of the communities are not well developed.[50] This can be attributed to the fact that the local people who are involved in the extraction of sand on a small scale use hand tools, and as such they cannot benefit as much as the Chinese and other foreign companies who use excavators and other machineries, which definitely require enormous amounts of energy such as diesel and petrol to operate. This makes access to modern energy essential for the local people to be able to utilise sand for urbanisation.[51]

In a nutshell therefore, access to advanced technology and machinery is essential in ensuring that artisanal miners in SSA (be it in gold, sand, or salt mining) enjoy decent working conditions, which is one of the SDGs. Access to modern energy is key in operating the suggested advanced technology. Additionally, these activities are capable of improving the livelihood of various people in SSA, reducing poverty, and ensuring economic growth—all of which are SSA targets set to be met by UN member states by 2030.

---

[49] See the Guardian, https://www.theguardian.com/cities/2017/feb/27/sand-mining-global-environmental-crisis-never-heard, last visited on 14 January 2018.

[50] For a full discussion on this and also to watch interviews from the local people where sand is extracted from in Uganda, see Victoria Nalule YouTube Channel, "Preserve Sand in Africa", https://www.youtube.com/watch?v=TfEU7w7GoZ0&t=143s, last visited on 14 January 2018. In this video, I also highlight and show the kind of infrastructure around these sand resource rich areas in Uganda.

[51] One important issue to note here is that the environmental impacts left behind by some of these Chinese companies are massive. To get an idea of some of these impacts, see Victoria Nalule YouTube Channel, "Mining Closures in Africa", https://www.youtube.com/watch?v=u5t0_n7Gr5w&t=40s, last visited on 14 January 2018. This video also benefited from the field work I carried out in Uganda. See also scholarly articles on the subject: Sreebha, S. and Padmalal, D., 2011. Environmental impact assessment of sand mining from the small catchment rivers in the southwestern coast of India: A case study. *Environmental Management, 47*(1), pp. 130–140; de Leeuw, J., Shankman, D., Wu, G., de Boer, W.F., Burnham, J., He, Q., Yesou, H. and Xiao, J., 2010. Strategic assessment of the magnitude and impacts of sand mining in Poyang Lake, China. *Regional Environmental Change, 10*(2), pp. 95–102.

The above discussion clearly brings out the reasons as to why energy poverty/energy access is a major problem in SSA. This book therefore aims at contributing to the ongoing efforts to tackle the challenge of energy access in SSA, specifically by exploring critically how regional organisations can work more effectively to promote cooperation in this area. In doing so, it attempts to make a contribution to our understanding of the notion of energy access itself.

1.3 Outline of the Book

In this book, a six-step framework is developed over seven chapters. Chapter 2 discusses the notion of energy access and also outlines the existing initiatives to tackle the challenge in SSA. The chapter specifically explores how the problem of energy access is understood by different organisations. In this regard, the different definitions of energy access are highlighted together with the various steps taken by different international organisations to address this problem.

Chapter 3 discusses the evolution of regionalism and the notion of regional cooperation as a means of tackling energy access challenges in SSA. The different forms of regional cooperation mechanisms are set out and distinctions are determined from existing mechanisms, with failed previous attempts by the various SSA regional organisations such as EAC to achieve regional integration highlighted to determine if the same mistakes are likely to affect the ongoing integration process and regional efforts to tackle energy access. This chapter also helps in illustrating the different regional cooperation mechanisms that will be the focus of this book and how these can be evaluated. The SSA energy access challenges are also explored, and an analysis of how these could be handled at the regional level is presented.

Chapter 4 looks at the role of regional cooperation in the development of renewable energy as a means of tackling both energy access and climate change, specifically examining the nexus between energy access, renewable energy, and climate change. In this regard, existing renewable energy regional mechanisms in SSA are discussed. Further, the Paris Agreement as a global instrument in response to climate change is examined to ascertain if its provisions provide room for regional efforts to tackle climate change and energy access in SSA.

The fifth chapter discusses regional energy infrastructure and examines how regional cooperation mechanisms are employed by EAC, SADC, and ECOWAS. Chapter 5 primarily looks at the role of these organisations in the establishment and development of regional energy infrastructure as one of the mechanisms of addressing energy access challenges in SSA.

In Chap. 6, a comparative analysis is employed to discuss national regulatory authorities in the various countries of SADC. The chapter then examines efforts by these countries to harmonise and have common regulatory rules at a regional level. Moreover, given the fact that most SSA countries and regional organisations face similar energy challenges, the chapter briefly looks at the experience of the European Union (EU) and how EU regional efforts in harmonising energy regulations can be of relevance and provide lessons to SSA regional organisations.

Chapter 7 sets out the concluding remarks, answering the questions raised and provides recommendations for policy makers and other relevant stakeholders.

References

Abu-Ghaida, D., & Klasen, S. (2004). The costs of missing the Millennium Development Goal on gender equity. *World Development, 32*(7), 1075–1107.

Ansell, N. (2017). Achieving gender parity in education: Achievements and limitations of Millennium Development Goal 3. In *Laboring and learning* (pp. 111–130). Springer.

Assessment, U.G.M. (2013). *Sources, emissions, releases and environmental transport* (p. 42). Geneva, Switzerland: UNEP Chemicals Branch.

Attaran, A. (2005). An immeasurable crisis? A criticism of the Millennium Development Goals and why they cannot be measured. *PLoS Medicine, 2*(10), 318.

Camacho-Poyato, E., & Montesinos, P. (2011). The paradox of irrigation scheme modernization: More efficient water use linked to higher energy demand. *Spanish Journal of Agricultural Research, 9*(4), 1000–1008.

Cecelski, E., & Policy, A. A. E. (2002). Enabling equitable access to rural electrification: Current thinking and major activities in energy, poverty and gender. *World Development Report, 1*, 2–3.

Danielsen, K. (2012). *Gender equality, women's rights and access to energy services*. A paper presented to the Ministry of Foreign Affairs of Denmark.

East African Community Treaty, 1999.

Easterly, W. (2009). How the millennium development goals are unfair to Africa. *World Development, 37*(1), 26–35.

Horton, R. (2014). Offline: Why the sustainable development goals will fail. *The Lancet, 383*(9936), 2196.

International Energy Agency: World Outlook Special Report, 2017.

Kirabira, J. B., Kasedde, H., & Semukuuttu, D. (2013). Towards the improvement of salt extraction at Lake Katwe. *International Journal of Scientific & Technology Research, 2*(1), 76–81.

Langarita, R., Chóliz, J. S., Sarasa, C., Duarte, R., & Jiménez, S. (2017). Electricity costs in irrigated agriculture: A case study for an irrigation scheme in Spain. *Renewable and Sustainable Energy Reviews, 68*, 1008–1019.

Lecina, S., Isidoro, D., Playán, E., & Aragüés, R. (2010). Irrigation modernization and water conservation in Spain: The case of Riegos del Alto Aragón. *Agricultural Water Management, 97*(10), 1663–1675; Rodríguez-Díaz, J. A., Pérez-Urrestarazu, L.

Rahman, A., & Bhatt, B. P. (2017). Scope and options of solar energy use in agriculture in eastern region of India. *Journal of AgriSearch, 4*(1), 76–79.

S.H. (2007). *Gender, generation and poverty: Exploring the feminization of poverty in Africa, Asia and Latin America*. Edward Elgar Publishing.

Sachs, J. (2005). *The end of poverty: How we can make it happen in our lifetime*. Penguin UK.

SADC Energy Monitor, 2016.

Sreebha, S., & Padmalal, D. (2011). Environmental impact assessment of sand mining from the small catchment rivers in the southwestern coast of India: A case study. *Environmental Management, 47*(1), 130–140; de Leeuw, J., Shankman, D.

Stafford-Smith, M., Griggs, D., Gaffney, O., Ullah, F., Reyers, B., Kanie, N., et al. (2017). Integration: The key to implementing the Sustainable Development Goals. *Sustainability Science, 12*(6), 911–919.

The 1992 SADC Treaty.

United Nations Sustainable Development Goals, 2016.

Veiga, M. M., Angeloci-Santos, G., & Meech, J. A. (2014). Review of barriers to reduce mercury use in artisanal gold mining. *The Extractive Industries and Society, 1*(2), 351–361.

Warschauer, M., & Ames, M. (2010). Can one laptop per child save the world's poor? *Journal of International Affairs, 64*(1), 33–51.

Wu, G., de Boer, W. F., Burnham, J., He, Q., Yesou, H., & Xiao, J. (2010). Strategic assessment of the magnitude and impacts of sand mining in Poyang Lake, China. *Regional Environmental Change, 10*(2), 95–102.

2

Energy Access in Sub-Saharan Africa

2.1 Understanding the Concept of Energy Access

Energy access remains a major priority to be addressed at the national, regional, and international levels, and this requires not only more effective mobilisation and use of domestic and external funding, but also the development and implementation of innovative policy frameworks.[1] Moreover, this cannot be achieved at once but rather it is a progressive process.

Energy is central to issues of development, global security, and environment and as such it has been put at the forefront of the global agenda. Industrialisation is escalating in developing countries with particular relevance for the agricultural sector, which the majority of the people in developing countries rely on. The agricultural sector is becoming increasingly modernised and thus increasing the demand for energy in these countries.[2] Moreover, as noted in Chap. 1, energy is also essential for the

[1] Brew-Hammond, A., 2010. Energy access in Africa: Challenges ahead. *Energy Policy, 38*(5), pp. 2291–2301.

[2] Aglina, M.K., Agbejule, A. and Nyamuame, G.Y., 2016. Policy framework on energy access and key development indicators: ECOWAS interventions and the case of Ghana. *Energy Policy, 97*, pp. 332–342.

V. R. Nalule, *Energy Poverty and Access Challenges in Sub-Saharan Africa*, Energy, Climate and the Environment, https://doi.org/10.1007/978-3-319-95402-8_2

provision of basic services such as health care and education; it is also a prerequisite to sustainable development and essential in the global efforts to eradicate poverty. Despite this, a substantial proportion of people in developing countries still lack access to modern energy. The actual percentage of population with access to modern energy varies significantly between countries and across regions, and this is partly due to the nature of national socio-economic situations and energy resource availability in differing settings.[3]

It is estimated that approximately 1.2 billion people worldwide have no access to modern energy such as electricity and nearly 3 billion people rely on traditional biomass (such as wood and charcoal) for cooking and heating.[4] SSA is one of the regions still struggling with the challenge of energy access with over 290 million people having no access to modern energy such as electricity.[5] This is despite the region's richness in energy resources with an estimated 65 billion barrels of proven oil reserves, equivalent to around 5% of the world total.[6] According to the African Development Bank (AfDB), Africa's power connectivity stands at 39 MW per million inhabitants, the lowest for any developing region.[7] Besides having the lowest level of connectivity in the region, recurrent outages and load shedding are also a major challenge.[8]

2.1.1 Energy Access Definitions

As noted in Chap. 1, energy access has become a focal topic in the twenty-first century and it remains a major challenge which is yet to be overcome

[3] Uddin, N. and Taplin, R., 2015. Regional cooperation in widening energy access and also mitigating climate change: Current programs and future potential. *Global Environmental Change, 35*, pp. 497–504.

[4] United Nations Foundation: Achieving Universal Energy Access, http://www.unfoundation.org/what-we-do/issues/energy-and-climate/clean-energy-development.html, last visited on 20 November 2015.

[5] International Energy Agency: World Energy Outlook Special Report, 2014. Page 13.

[6] Ibid. Page 14.

[7] African Development Bank Group: PIDA, http://www.afdb.org/fileadmin/uploads/afdb/Documents/Generic-Documents/PIDA%20brief%20Energy.pdf, last visited on 25 November 2015.

[8] The AFDB estimates that more than 30 African countries experience recurrent outages, with opportunity costs amounting to as much as 2% of the total annual value of the economy.

[9] See International Energy Agency: World Energy Outlook, Special Report, 2017. Page 11.

in developing regions such as Africa and Asia. Besides the countries in Africa, developing countries in Asia are also faced with the challenge of energy poverty. However, Asia has made some significant progress as the electrification rate in the region increased to 89% in 2016, up from 67% in 2000. Comparatively, the electrification rate in SSA is only 43% making it the least electrified region.[9]

Despite the urgent need to tackle energy access, the problem continues to persist especially in Africa, and this has been attributed to the fact that energy access was not given international attention in the past as compared to other development topics such as human sanitation and access to clean water. In this regard, access to energy received much less interest and limited investment and political backing.[10] Moreover, the limited global attention given to energy access in the past has resulted in limited efforts to come up with an agreed definition of the concept; in this regard, there is no internationally agreed definition of energy access, and as such it has been understood differently by different scholars, and regional and international organisations.

Although the concept of energy access has only recently gained academic momentum, different terms with almost the same effect were highlighted as far back as the 1970s, with discussion on issues such as shortage of firewood in developing countries[11] and renewable energy, particularly solar energy, being tackled.[12] The connection between topics discussed in the 1970s such as the use of firewood for energy purposes with the current concept of energy access relates to the fact that firewood is now considered as a traditional form of energy; in the recent definitions of energy access, there is an emphasis for the need to shift from traditional to modern energy.[13]

It can therefore be argued that the concept of energy access was not alive in the twentieth century and the centuries before. It is only in the twenty-first century that the concept has gained momentum with

[10] Uddin, N. and Taplin, R., 2015. Regional cooperation in widening energy access and also mitigating climate change: Current programs and future potential. *Global Environmental Change, 35*, pp. 497–504.

[11] Eckholm, E., 1975. *The other energy crisis: Firewood*. Worldwatch Paper, (1).

[12] Gergacz, J.W., 1979. Solar energy law: Easements of access to sunlight. *New Mexico Law Review, 10*, p. 121.

various landmarks in the literature on the topic resulting from this. Some have suggested the need to embrace renewable energy as a form of tackling energy access,[14] while others have examined the impact of power sector reforms on the poor people in relation to energy access.[15] Although the literature on energy access has been growing steadily, not much attention has yet been given to defining the concept and indeed scholars have focused the definition by distinguishing between modern energy and traditional energy.[16] Additionally, it is worth noting that energy access is also referred to as energy poverty in some literature.[17]

Moreover, there is a high reliance on how inter-governmental organisations such as the UN and IEA have defined the concept of energy access, and their approach is highly relied on by different scholars.[18] This therefore shows the influence these international organisations have on global academia and underlines the importance of such organisations framing the problems or concepts correctly from the beginning.

Taking all this into consideration, we shall explore how some inter-governmental organisations and international organisations have endeavoured to define energy access.

---

[13] Modern energy also includes renewables such as solar energy, and traditional energy is often taken to mean traditional biomass, which definitely includes the reliance on firewood, mostly in rural areas.

[14] Painuly, J.P., 2001. Barriers to renewable energy penetration: A framework for analysis. *Renewable Energy, 24*(1), pp. 73–89.

[15] Karekezi, S., Kimani, J., Mutiga, A. and Amenya, S., 2003. Energy services for the poor in Eastern Africa: Sub-regional "Energy Access" study of East Africa. A report prepared for "Energy Access" working group of the Global Network on Energy for Sustainable Development.

[16] Karekezi, S., Kimani, J. and Onguru, O., 2008. Energy access among the urban poor in Kenya. *Energy for Sustainable Development, 12*(4), pp. 38–48.

[17] Bazilian, M., Nussbaumer, P., Cabraal, A., Centurelli, R., Detchon, R., Gielen, D., Rogner, H., Howells, M., McMahon, H., Modi, V. and Nakicenovic, N., 2010. Measuring energy access: Supporting a global target. Earth Institute, Columbia University, New York, USA.

[18] Bazilian, M., Nussbaumer, P., Cabraal, A., Centurelli, R., Detchon, R., Gielen, D., Rogner, H., Howells, M., McMahon, H., Modi, V. and Nakicenovic, N., 2010. Measuring energy access: Supporting a global target. Earth Institute, Columbia University, New York, USA. In this report, not much effort was made to define energy access rather the concept of energy poverty was defined relying mostly on how inter-governmental organisations, such as the IEA, have defined energy access.

2.1.1.1 2010 Definition by the United Nations

In 2010, the UN endeavoured to define energy access as "access *to a basic minimum threshold of modern energy services for both consumption and productive uses*".[19] This definition points to the use of energy not only for household consumption but other productive sectors such as businesses. This definition will be analysed later on in this chapter.

2.1.1.2 2014 Definition by International Energy Agency

In 2014, the *International Energy Agency (IEA)* defined energy access by directly connoting it to the term "modern", thus defining modern energy access as "*household having access to electricity and to a relatively clean, safe, means of cooking*".[20] In this definition the quality and quantity of electricity are considered, as elaborated below:

According to the IEA energy modelling results, 250 kilowatt-hours (kWh) per year is assumed to be the electricity consumption for rural households, while 500 kWh is for urban households. Both are calculated based on an assumption of five people per household, this assumed electricity consumption is also influenced by the various purposes of electricity.[21] Recognizing the fact that traditional biomass is mostly relied on in rural areas for cooking, the IEA stretches the definition of modern energy access, to also include access to cooking facilities that are considered safer, more efficient and more environmentally sustainable than the traditional facilities that make use of solid biomass which is common practice across sub-Saharan Africa.[22]

---

[19] The Secretary General's Advisory Group on Energy and Climate Change: Energy for a Sustainable Future, Report and Recommendations, 28 April 2010, New York. See http://www.un.org/millenniumgoals/pdf/AGECCsummaryreport[1].pdf, last visited on 17 November 2015.

[20] International Energy Agency: World Energy Outlook-Methodology for Energy Access Analysis, 2014 www.worldenergyoutlook.org/media/weowebsite/EnergyAccess_Methodology_2014.pdf, last visited on 6 October 2015.

[21] For instance, whereas in rural areas it might be used to mostly provide light, in urban areas it might stretch to refrigeration uses.

[22] International Energy Agency: World Energy Outlook-Methodology for Energy Access Analysis, 2014 www.worldenergyoutlook.org/media/weowebsite/EnergyAccess_Methodology_2014.pdf

The above IEA definition brings into play aspects of quality, quantity, and use of energy. It also highlights the different areas including rural and urban and how these consume energy differently. The UN definition leaves the aspect of modern energy hanging and this therefore makes the definition ambiguous while the IEA definition on the other hand generally defines modern energy access, but it is also ambiguous with regard to the aspect of "relatively clean, safe, means of cooking". Nonetheless, the term "modern" is generally introduced in the access discourse to make a distinction between traditional forms of energy like firewood or agricultural residues and commercial forms of energy like electricity or liquefied petroleum gas (LPG).[23]

2.1.1.3 2015 Definition by the World Bank

In 2015, the World Bank endeavoured to redefine energy access.[24] In the report entitled "*Beyond Connections Energy Access redefined*", the authors conceptualised a new multi-tier framework for defining and measuring access to energy. An argument brought forward in this report is that "binary metrics such as whether a household has an electricity connection, and whether a household cooks with non-solid fuels do not help understand the phenomenon of energy access. The report thus suggests a new definition and a metric of energy access that is broader, to cover energy for households, productive engagements and community facilities and focuses on the quality of energy being accessed."[25]

Though this report also suggests widening the mechanisms of tackling the challenge of energy access, this definition seems to be very ambiguous. There is a need to precisely understand what the problem is and find a precise solution. Moreover, the report suggests that the access challenge

---

[23] Abeeku Brew-Hammond: Challenges to increasing Access to Modern Energy Services in Africa, codlogics_ff_201312 30_1388442782.report on West African energy ministers conference 2007.

[24] Bhatia, M. and Angelou, N., 2015. *Beyond connections energy access redefined.* Energy Sector Management Assistance Program, World Bank.

[25] Bhatia, M. and Angelou, N., 2015. *Beyond connections energy access redefined.* Energy Sector Management Assistance Program, World Bank.

should not only be limited to the people without electricity connections but rather to the households with poor and unreliable electricity supplies.[26] This seems to be a valid argument considering the common elements of energy access, which include accessibility, reliability, affordability, and sustainability. However, the problem with this expansion is that it diverges away from the focal problem in SSA which is mainly one of accessibility. Of course, other elements are also important but we need to narrow down the problem to find a precise solution and perhaps identify other elements that can also be tackled separately under the auspices of a different concept, thus finding specific and precise solutions. For instance, the element of affordability in relation to energy can be tackled separately, and not necessarily mixing it with accessibility and reliability. This is simply due to the fact that energy can be accessible, reliable, and sustainable but nevertheless not affordable to the people—especially the people in rural areas.

2.1.1.4 2017 Definition by the International Energy Agency

In their recent 2017 report, the IEA defined energy access as "*a household having reliable and affordable access to both clean cooking facilities and to electricity, which is enough to supply a basic bundle of energy services initially, and then an increasing level of electricity over time to reach the regional average*".[27]

The above IEA definition focuses mostly on households and the question arises as to whether productive sectors such as industries and transport do not need energy? Additionally, there is a mix in the issues

[26] Bhatia, M. and Angelou, N., 2015. *Beyond connections energy access redefined*. Energy Sector Management Assistance Program, World Bank.

[27] In this report, a basic bundle of energy services means, at a minimum, several light bulbs, task lighting (such as a flashlight), phone charging, and a radio. Access to clean cooking facilities means access to (and primary use of) modern fuels and technologies, including natural gas, liquefied petroleum gas (LPG), electricity, and biogas, or improved biomass cook stoves (ICS, as opposed to the basic biomass cook stoves and three-stone fires used in developing countries. See International Energy Agency: World Energy Outlook, Special Report, 2017. Page 21. It can be accessed at https://www.iea.org/publications/freepublications/publication/WEO2017SpecialReport_EnergyAccessOutlook.pdf, last accessed on 12 March 2018.

concerning accessibility and affordability, which have to be separated while tackling energy poverty challenges in SSA, an argument which will be maintained in this book.

2.1.1.5 Analysis of Energy Access Definitions and Suggested New Definition

Analysing the definitions above, it can be noted that energy access does not necessarily mean access to electricity. However, if the electricity accessed can provide lighting and also be used for cooking then this definition which connotes "energy access to mean access to electricity" can be acceptable to some extent, especially with regard to households. Electricity is surely pivotal in defining energy access, but what about access to other forms of energy such as petroleum products like diesel and petro—which are key for industrialisation and the transport sector? Does it mean that developing countries do not care about industrialisation and transportation? Well these are some of the questions we have to put into consideration while trying to define energy access.

From the literature review, we also note that there is an ongoing struggle to arrive at a universally agreed definition of energy access. This is problematic as it implies that the mechanisms taken to solve the challenge are likely to be misconceived and delayed, because misconceived concepts will mostly lead to misconceived solutions. Moreover, in spite of all the international initiatives to tackle the challenge, if the problem cannot be framed correctly from the beginning, then the challenge of energy poverty is likely to persist in SSA. Additionally, new initiatives to tackle the problem will also continue to rely and be based on such misconceived concepts, thus delaying the progress to successfully tackle energy access in developing countries.[28]

[28] Examples of new initiatives that are relying on the concept include the G20 through their energy access action plan. This is fully covered in Chap. 1 under the heading of Existing Initiatives.

[29] See International Energy Agency: World Energy Outlook, Special Report, 2017. Page 22. It can be accessed at https://www.iea.org/publications/freepublications/publication/WEO2017SpecialReport_EnergyAccessOutlook.pdf, last accessed on 12 March 2018.

Considering that households, social institutions such as schools and hospitals, business enterprises such as shop keeping and blacksmiths' forges may require energy for different purposes and in different quantities, it becomes relevant to separate the definition depending on the demand and use of energy in these different sectors.

Furthermore, on the issue of affordability, we have to be aware that people in urban areas are most likely to afford different quality of energy as compared to rural areas, for instance LPG or electricity for cooking maybe more suitable and with more markets in urban areas than in rural areas. In defining energy access, it is also necessary to consider the kind of technology needed, be it grid, mini-grid, and off-grid solutions for electricity and advanced gas (LPG), stove, and biogas systems for cooking. There is a general acceptance that there is a shift from solid biomass (firewood, charcoal) to liquid fuels (ethanol, kerosene) or gaseous fuels (LPG, natural gas) or electricity for cooking and the connection to the national grid as threshold for access.

In this book, a proposed definition of energy access is "*households, social institutions and industries having access to electricity and other forms of modern energy*" in SSA. The definition concentrates on households because they are the ones mostly affected by in-house air pollution. In this definition, a three-phase energy access framework is however suggested: the first phase should concentrate on households accessing clean energy for lighting and cooking; the second phase should concentrate on energy being affordable by the people and later being reliable; the third phase should concentrate on the extension of energy to businesses, industries, and social institutions such as health care services and schools.

The suggested definition also assumes the separation of the elements of accessibility, reliability, and affordability. This separation is essential because these elements are different and require different solutions. Mixing them up means that the initiatives taken to solve the challenge will not be effective. Whereas the issue of accessibility requires the establishment of hard infrastructure such as electricity transmission lines, setting up favourable legal frameworks to attract the needed investments in the energy sector and other aspects, on the other hand, the issues concerning affordability have much to do with the general economic situation in the country.

In some developing countries where the majority of people are depending on one dollar a day, it makes no sense to mix accessibility and affordability because at the end of the day they will not be able to afford the electricity even if it was accessible. There are many people in the developing countries who can afford modern energy but cannot access it. For instance, it has been noted by IEA that in SSA, at least 120 million people are living above the poverty line but are without electricity, and the situation is the same in developing Asia, with 70 million people above poverty line but without electricity.[29] Additionally, the issue of "fuel stacking" has also been highlighted as a barrier to achieving universal access to modern energy. Fuel stacking is a situation where a household gains access to affordable modern fuels such as electricity and LPG but continues to use biomass (with regard to fuel stacking, the IEA refers to the example of the Hubei Province in China, where a survey of rural household fuel use found that 99% of households used at least two types of fuels and that fewer than 10% of households had abandoned the use of biomass altogether when they could afford to do so).[30]

Taking stock of the above, the argument in this book therefore is that there is a need to handle the elements of energy access differently and separately so that we can find specific solutions to the specific challenges. Generalising these concepts will only lead to misconceived solutions and we shall forever sing the song of "energy poverty/energy access in SSA".

Additionally, we should not ignore other forms of energy like petroleum products, which are essential for industrialisation and transportation. This book will not dwell on this subject, though its argument implies that these should also be part of energy access.

2.1.2 Similar Terms to Energy Access

2.1.2.1 Energy Poverty

Energy poverty is defined as the absence of sufficient choice in accessing adequate, affordable, reliable, quality, safe, and environmentally benign

[30] Ibid.

energy services to support economic and human development.[31] It is also argued that energy poverty can take different forms including a lack of access to modern energy services, a lack of reliability, and concerns about affordability of access.[32] The energy poverty term is mostly interchanged with that of energy access, and as such both terms will be used interchangeably in this book.

It is imperative to note that although this book focuses on energy poverty in the developing countries, specifically those in SSA, energy poverty has also been documented as a challenge in the developed countries. Recent literature points to the fact that even countries which have acquired universal and reliable access to modern energy still encounter affordability challenges.[33] The IEA, based on the methodology of the spending measure[34] and on survey responses on whether households are able to keep their homes sufficiently warm, estimates that over 15% of the total population in developed economies suffers from energy poverty.[35]

In this regard, looking at the issue of affordability, which limits or makes it hard for people in the developed countries to enjoy the already accessible and reliable energy services, we can conclude that energy

[31] See Cecelski, E., 2002. Enabling equitable access to rural electrification: Current thinking and major activities in energy, poverty and gender: Quoting World Energy Assessment 2000, Sept 1999 draft.

[32] International Energy Agency: World Energy Outlook, Special Report, 2017. Page 24. It can be accessed at https://www.iea.org/publications/freepublications/publication/WEO2017SpecialReport_EnergyAccessOutlook.pdf, last accessed on 12 March 2018.

[33] The IEA, referring to the UK Fuel Poverty Strategy in 2001 and the subsequent review in 2012, highlights three main reasons for energy poverty in developed countries, and these include low incomes, high energy costs, and energy inefficient dwellings. See International Energy Agency: World Energy Outlook, Special Report, 2017. Page 24. It can be accessed at https://www.iea.org/publications/freepublications/publication/WEO2017SpecialReport_EnergyAccessOutlook.pdf, last accessed on 12 March 2018.

[34] The spending measure methodology takes into account the proportion of households spending more than 10% of their income on energy. For a full discussion on this see International Energy Agency: World Energy Outlook, Special Report, 2017. Page 24. It can be accessed at https://www.iea.org/publications/freepublications/publication/WEO2017SpecialReport_EnergyAccessOutlook.pdf, last accessed on 12 March 2018.

[35] See International Energy Agency: World Energy Outlook, Special Report, 2017. Page 24. It can be accessed at https://www.iea.org/publications/freepublications/publication/WEO2017SpecialReport_EnergyAccessOutlook.pdf, last accessed on 12 March 2018.

poverty is indeed a global challenge which needs to be given more attention both in the developing and developed countries. This also emphasises the argument in the previous sections, which is to the effect that there is a need to redefine energy access/ energy poverty in the developing countries—specifically the need to tackle the aspects of accessibility, reliability, and affordability separately. As noted earlier, developed countries have successfully tackled the issue of accessibility to and reliability of modern energy services, and as such they are now only working towards addressing the issue of affordability. In developing countries however, the ambiguity of the definition of energy access/energy poverty applied by the international organisations leaves us confused about which aspect of energy poverty to handle first—be it accessibility, reliability, or affordability.

2.1.2.2 Energy Security

Whereas in the past energy security was primarily associated with oil supply, now with the increasing complexity of energy systems the term has expanded. The IEA in this regard defines energy security as "the uninterrupted availability of energy sources at an affordable price".[36]

This definition presumes a situation where a country already has access to modern and reliable energy services and as such the countries concerned are only concerned about ensuring that these energy services are constant and uninterrupted. In this regard, energy security is not a term to be used in developing countries which are still struggling with accessing energy services, but rather it suits best the developed countries which have ready energy infrastructure in place to ensure frequent access to reliable and affordable energy services. Additionally, this definition does not only look at energy for providing lighting and cooking, but also energy to ensure the continued industrialisation and functioning of the transport sector in the developed countries. Energy security therefore considers not

[36] International Energy Agency, https://www.iea.org/topics/energysecurity/subtopics/whatisenergysecurity/, last visited on 30 January 2018.

only access to electricity and liquefied natural gas for households, but also includes the continued supply of oil and other petroleum products for industries and other productive sectors.

2.2 Existing Initiatives to Tackle the Challenge of Energy Access in SSA

There are already initiatives in existence to tackle the challenge of energy access; however, the problem is still escalating in SSA. The active participation of various sectors at the regional, national, and international level has indeed been encouraged.

In recent years, we have seen an increase in several initiatives—bilateral, multilateral, and non-governmental initiatives—some of which work closely with SSA governments and regional bodies such as the African Union (AU), the African Development Bank, and the New Partnership for Africa's Development among others in order to tackle the challenge of energy access. These initiatives include among others Sustainable Energy for All (SE4ALL) initiative, Japan's power infrastructure support programme, the US Power Africa initiative, and the EnDev programme (Germany, the Netherlands, the United Kingdom, and others). Some of the existing initiatives are discussed here. There is however a need for strong coordination among these various initiatives to avoid unnecessary duplication and to be able to ensure a maximum degree of synergy.

2.2.1 Sustainable Energy for All (SE4ALL)

Energy access has gained a lot of attention following the adoption of the UN Sustainable Development Goals (SDG), which under goal 7 emphasises the access to affordable, reliable, sustainable, and modern energy for all.[37] It has been noted that electrification is one of the most difficult

[37] Goal 7 of the Sustainable Development Goals, http://www.un.org/sustainabledevelopment/energy/, last accessed on 4 October 2017.

aspects to achieve of the SDGs.[38] To give effect to the UN General Assembly Resolution 65/151 in 2011 titled the "International Year of Sustainable Energy for All", the SE4ALL initiative was launched by UN Secretary-General Ban Ki-moon and co-led by World Bank President Jim Kim. The establishment of SE4ALL has its roots in the recommendations of the April 2010 report published by the Secretary-General's Advisory Group on Energy and Climate Change (AGECC). SE4ALL details how governments, businesses, and civil society can work together to make sustainable energy for all a reality by 2030. The main objectives of SE4ALL include "*ensuring universal access to modern energy; doubling the rate of improvement in energy efficiency and doubling the share of renewable energy in the global energy mix*".[39]

Developments aimed at achieving the objectives of SE4ALL include the establishment of a global Energy Access Practitioner Network which was launched by the UN Foundation. The Network is focused on the removal of market barriers to the effective delivery of energy services by promoting the adoption of new technologies and innovative financial and business models. With regard to energy access in Africa, the SE4ALL African Hub was established with the mission of coordinating and facilitating the implementation of the SE4ALL initiative in the African continent, and the Hub is part of the SE4ALL global implementation structure and arrangements.[40] The AfDB holds the Hub Secretariat responsible for carrying out the activities of the Hub and its work programme. Guidance for the operations of the Hub Secretariat are provided by the Oversight and Operations Committee (OOC), which is comprised of one representative of the African Union Commission (AUC), the NEPAD planning and Coordinating Agency (NPCA), AFDB, the United Nations Development Program (UNDP), and a representative of the Regional Economic Communities (REC).

---

[38] Bazilian, S.T.M., 2017. The role of international institutions in fostering sub-Saharan Africa's electrification.

[39] Sustainable Energy for All: Objective. Retrieved at http://www.se4all.org/our-vision/our-objectives/, last visited on 2 November 2015.

[40] See Sustainable Energy for All: African Hub, http://www.se4all.org/hubs/african-hub/, last visited on 2 November 2015.

Another initiative of SE4All in Africa is the establishment of the African Energy Leaders Group (AELG) launched at Davos early in 2015. The AELG is a working group of African political and economic leaders at the highest level, with the aim of pooling their complementary skills to build momentum for a new vision and new solutions to energy challenges. The key goal of AELG is to provide access to reliable, affordable, and efficient energy to all Africans by 2030, which is in line with the SE4All initiative of ensuring universal access to modern energy services by 2030.[41] Further, there is the SE4All Knowledge Hub, hosted by the Energy Sector Management Assistance Program (ESMAP).[42] Work under the Knowledge Hub focuses on two areas: research that supports the implementation of SE4ALL initiatives through frameworks that track and monitor progress and provide feedback; and knowledge management work that collects information and data around key policy experiences and cases. The Hub has four main initiatives: Global Tracking Framework, The Readiness for Investment in Sustainable Energy, The Multi-Tier Framework for Measuring Access, and the State of the Energy Access Report.

2.2.2 The G20 Energy Access Action Plan: Voluntary Collaboration on Energy Access (the "Action Plan")

The commitment of the G20 leaders to ensure access to affordable and reliable energy for all is enshrined in the G20 framework of the Principles on Energy Collaboration, which were agreed upon at the Brisbane Summit. Basically, the G2O was established in 1999 to bring together industrialised and developing economies to discuss key issues facing the global economy, and it refers to the group of 20 Finance Ministers and Central Bank Governors of 19 countries including Argentina, Australia,

[41] Sustainable Energy for All: Sustainable Energy For All launches African Energy Leaders group, http://www.se4all.org/2015/01/23/sustainable-energy-launches-african-energy-leaders-group-davos/, last visited on 11 December 2015.

[42] See Sustainable Energy for All: Tracking Progress, http://www.se4all.org/tracking-progress/, last visited on 11 December 2015.

Brazil, Canada, China, France, Germany, India, Indonesia, Italy, Japan, Mexico, Russia, Saudi Arabia, South Africa, South Korea, Turkey, the United Kingdom, and the United States. The remaining seat is held by the EU, which is represented by the rotating Council presidency and the European Central Bank.

Energy Ministers from the G20 group on 2 October 2015 adopted an Energy Access Action Plan (the "Action Plan") in Istanbul Turkey, and this is intended to help fight energy poverty.[43] The Action Plan was drafted by SE4ALL in cooperation with 15 international organisations, which included the AFDB, African Union, World Bank, and the IEA. The first phase of the Action Plan is focused on improving electricity access in SSA, although the Action Plan is envisaged to be a multi-phase plan focusing on different regions and thematic issues. The Action Plan is intended to strengthen the collaboration of G20 members on energy access issues and to support existing initiatives at the regional, national, and international levels. In an endeavour to support the existing initiatives, G20 Members acknowledged the importance of providing sufficient resourcing to SE4ALL and other international organisations through voluntary contributions and in-kind contributions to assist work under the Action Plan to progress. The Global Facilitation Team of the SE4ALL initiative is intended to support collaboration under the Action Plan, cooperating with other relevant organisations to prepare regular input into the G20 Energy Sustainability Working Group (ESWG).

2.2.3 USAID Power Africa Program

This initiative was launched by President Obama in 2013, and its main aim is to increase access to electricity on the African continent. It intends to bring together technical and legal experts, the US private sector, SSA governments, and the AfDB in partnerships to increase Africans' access to power. The objectives of the initiative are to develop 60 million new

[43] G20 Ministers adopt Africa-focused Energy Access Action Plan, http://www.se4all.org/2015/10/07/g20-ministers-adopt-africa-focused-energy-access-action-plan/, last visited on 8 December 2015.

electricity connections and 30,000 MW of new generation capacity in all SSA countries.[44] Power Africa also recognises the need to bridge the financing gap by de-risking investments with the aim of attracting public and private energy investors.

2.2.4 Japan-Africa Energy Initiative (JAEI)

An agreement between the Government of Japan and the AfDB was signed on 3 July 2017 to develop the Japan-Africa Energy Initiative, which is intended to contribute to the ongoing efforts of achieving universal energy access on the African continent by 2025. Under the umbrella of this initiative, the Bank will lead the project development while the Japan government intends to provide up to USD 6 billion to support energy access projects.[45]

2.2.5 The EnDev Programme (Germany, the Netherlands, the United Kingdom, and Others)

The Energising Development (EnDev) programme aims at promoting access to modern energy in developing countries. It is a multi-donor partnership, supported by governments of the Netherlands, Germany, Norway, the United Kingdom, Switzerland, and Sweden.

2.2.6 Regional Initiatives

As will be discussed later, there are also regional initiatives to tackle the problem of energy access; however as noted by the IEA, a significant gap exists in practice between the potential gains from regional cooperation in SSA and the actual record of achievement.[46] The same argument can

[44] USAID Power Africa Programme.

[45] https://www.afdb.org/en/news-and-events/afdb-and-government-of-japan-launch-japan-africa-energy-initiative-17154/, last visited on 22 November 2017.

[46] International Energy Agency: World Energy Outlook Special Report, 2014. Page 179.

hold true for international initiatives; there are several of these but the challenge of energy access is escalating in SSA. The actual benefits international initiatives yield are trivial as compared to the efforts they put in, and reversing these slow results would require combined efforts and an agreed methodology and accountability in these initiatives. Moreover, lack of coordination in the various international initiatives has further played a non-trivial role in undermining efforts to tackle the challenge of energy access in SSA.

References

Abeeku Brew-Hammond: Challenges to increasing Access to Modern Energy Services in Africa, codlogics_ff_201312 30_1388442782.report on West African energy ministers conference 2007.

Aglina, M. K., Agbejule, A., & Nyamuame, G. Y. (2016). Policy framework on energy access and key development indicators: ECOWAS interventions and the case of Ghana. *Energy Policy, 97*, 332–342.

Bazilian, M., Nussbaumer, P., Cabraal, A., Centurelli, R., Detchon, R., Gielen, D., et al. (2010). Measuring energy access: Supporting a global target. Earth Institute, Columbia University, New York.

Bhatia, M., & Angelou, N. (2015). *Beyond connections energy access redefined.* Energy Sector Management Assistance Program, World Bank.

Brew-Hammond, A. (2010). Energy access in Africa: Challenges ahead. *Energy Policy, 38*(5), 2291–2301.

Cecelski, E. (2002). Enabling equitable access to rural electrification: Current thinking and major activities in energy, poverty and gender: Quoting World Energy Assessment 2000, Sept 1999 draft.

Eckholm, E. (1975). *The other energy crisis: Firewood.* Worldwatch Paper (1).

Gergacz, J. W. (1979). Solar energy law: Easements of access to sunlight. *New Mexico Law Review, 10*, 121.

International Energy Agency: World Energy Outlook Special Report, 2014.

International Energy Agency: World Energy Outlook, Special Report, 2017.

Karekezi, S., Kimani, J., & Onguru, O. (2008). Energy access among the urban poor in Kenya. *Energy for Sustainable Development, 12*(4), 38–48.

Oil & Gas Journal, as of January 1, 2013; EIA, U.S. natural gas reserves, 2010.

Painuly, J. P. (2001). Barriers to renewable energy penetration: A framework for analysis. *Renewable Energy, 24*(1), 73–89.

The Secretary General's Advisory Group on Energy and Climate Change: Energy for a Sustainable Future, Report and Recommendations, 28 April 2010, New York.

Uddin, N., & Taplin, R. (2015). Regional cooperation in widening energy access and also mitigating climate change: Current programs and future potential. *Global Environmental Change, 35*, 497–504.

3

Regionalism in Addressing Energy Access Challenges

The aim of this chapter is to begin to explore the concept of regionalism as a mechanism to address the energy challenges faced by the SSA nations. Prior to discussing the evolution of regionalism and the way in which the concept emerged in the SSA energy sectors it is imperative to develop a picture of the energy situation in SSA at the national level. This will give us an idea of the common challenges these countries face in their national energy sectors and how common regional mechanisms can be employed to solve the energy challenges.

Recognising the fact that most regions in SSA face similar energy challenges including unreliable electricity, reliance on traditional biomass to mention but a few, and taking into consideration that this book focuses on three specific SSA case studies (EAC, ECOWAS, and SADC), it is worth acknowledging the difficulty in discussing the national energy challenges facing all the countries in the three regional organisations. In this regard, then, the 15 countries that make up SADC will be briefly discussed to provide an overview of the key energy challenges they face at a national level together with the legal frameworks governing their energy sectors. The argument for this selection is that it will set out the basis for regional common efforts to tackle common energy challenges.

V. R. Nalule, *Energy Poverty and Access Challenges in Sub-Saharan Africa*, Energy, Climate and the Environment, https://doi.org/10.1007/978-3-319-95402-8_3

3.1 An Overview of the SSA Energy Situation at the National Level

3.1.1 Republic of Angola

Political stability has been maintained in Angola since the end of the civil war in 2002, and the country has witnessed both political and economic progress. Angola is the largest oil producer in SADC and second in Sub-Saharan Africa, behind Nigeria. The country holds 9 billion barrels of proven crude oil reserves, and natural gas reserves are estimated at 9.7 Tcf.[1]

The oil and gas sector is the main driver of the Angolan economy. It is estimated that oil represents 80% of government revenues; however, the weak performance of the oil sector internationally has had a negative impact on the country's economy.[2] Despite its richness in energy resources like oil, access to electricity is still very low in the country with only around 30% of the country's 18 million inhabitants having access to electricity, the majority of which is mostly found in urban areas. As such, the use of traditional biomass in the form of fuel wood is particularly prevalent in rural areas. This is attributed among others to the lack of adequate infrastructures for generation, transmission, and distribution.[3]

In order to attract private investors in the electricity sector and to ensure rural electrification in Angola, in July 2015 the parliament passed the General Electricity Law, resulting in the National Electricity Company (ENE) dismantled into three other firms: Public Electricity Production Enterprise EP (PRODEL) responsible for producing power; the National Transportation Network (RNT) responsible for transportation and substations; and Empresa de Electricidade de Luanda/Electricity Company

---

[1] See *Oil and Gas Journal*, http://www.ogj.com/articles/print/volume-112/issue-1/drilling-production/worldwide-look-at-reserves-and-production.html, last visited on 1 October 2015.

[2] The World Bank estimates that Angola's GDP declined from 6.8% in 2013 to 4.2% in 2014, and this is mostly attributed to the weak performance of the oil sector, which has contracted since 2011. World Bank, Angola Overview, retrieved from http://www.worldbank.org/en/country/angola/overview, last visited on 2 February 2016.

[3] Government of Angola and UNDP: Sustainable Energy For All: Rapid Assessment Gap Analysis, Angola, September 2015.

of Luanda (ENDE) responsible for distributing electricity throughout the country.

Angola is also open to FDI, having recently enacted a new private investment law No. 14/15 that established the principles and rules of access to benefits, incentives, and guarantees for investment in Angola. These incentives include tax deductions, levies, customs benefits, protection against expropriation, and compensation.[4] The Government of Angola also created a new investment agency, the *Agência para a Promoção do Investimento e Exportações de Angola/Agency for Investment Promotion and Export Angolain* (APIEX). Furthermore, Angola is a member of different multilateral and bilateral agreements, including among others the Multilateral Investment Guarantee Agency (MIGA), which provides private sector investment and dispute settlement assistance; the World Intellectual Property Organization (WIPO); and the SADC free trade protocol that aims to harmonise trade and customs regimes and reduce tariffs among the SADC member states.

3.1.2 Republic of Botswana

Botswana is landlocked and one of the most sparsely populated countries in the world. The country has experienced great economic progress since independence,[5] moving into the ranks of upper middle-income status, with an average annual growth rate of about 9%.[6] Minerals have played a big role in the economic success, especially the development of the diamond extraction industry, and the successful management of the revenue this has brought together with the prudent fiscal policies.[7]

Access to electricity in the country stands at 66% with the target to reach 80% by 2016.[8] The Electricity Supply Act regulates the sector and

---

[4] See Article 1, 20, 21, 22, 28 of the private investment law n. 14/15 August 11.

[5] The country gained independence from the British on 30 September 1966.

[6] World Bank: Botswana, Country Profile, http://web.worldbank.org/archive/website01321/WEB/0__CONTE.HTM, last visited on 2 February 2015.

[7] *Africa Energy Outlook: A focus on energy prospects in sub-Saharan Africa*, 2014. Page 178.

[8] See Ministry of Mines and Energy, Department of Energy, http://www.energy.gov.bw/?id=7852, last visited on 2 October 2015.

provides for licenses to be issued by the Minister for the generation and supply of electricity.[9] The electricity sector is dominated by the utility Botswana Power Corporation (BPC), under the Botswana Power Corporation Act (BPC Act).[10] With more investments seen as essential in the electricity sector, the Electricity Supply Act was amended in 2007 to enable private investors to participate in energy generation.[11] This reform liberalised the electricity market allowing the creation and licencing of independent producers and suppliers of electricity, an essential measure conducive to the development of the electricity supply industry.[12]

Botswana has several mechanisms in place to attract FDI. The Constitution under Article 8 provides protection from deprivation of property except in certain circumstances such as public interest and payment of prompt and adequate compensation.[13] The same protection is provided for under the Acquisition of Property Act.[14]

The Botswana Investment and Trade Center (BITC) is the government agency dedicated to promoting investment in the country. It is a merger between Botswana Export Development and Investment Authority (BEIDA) and the International Financial Sector (IFSC). BITC is an autonomous private sector led organisation mandated by an Act of Parliament to encourage and promote investments in the country.

The country is also a member state to the ICSID Convention, the 1958 New York Convention, the World Trade Organization (WTO), The African Caribbean Pacific-European Union Partnership Agreement (ACP-EU), and the Southern African Customs Union (SACU).

3.1.3 The Democratic Republic of Congo

The Democratic Republic of Congo (DRC) is the second largest country in Africa in terms of size and the eleventh largest in the world. The

[9] Electricity Supply Amendment Act, Chapter 73:01.

[10] Botswana Power Corporation Act, 1970 (Chapter 74:01).

[11] See article 3, 4, 5 of the Electricity Supply (Amendment) Act, 2007, available at http://www.gov.bw/Global/MMWER/Act%20No%20%2020%20AMMNEDMENT%20ESA.pdf

[12] Botswana National Development Policy 10. Page 238.

[13] The Constitution of Botswana, 1966.

[14] Section 17 of the Acquisition of Property Act, Chapter 32:10.

country has an estimated GDP of USD 32.9 billion and in 2014 the economic growth rate was estimated at 9%.

Natural resources such as diamonds, gold, copper, cobalt, and oil among others have contributed greatly to the economic success of the DRC. There is great potential for hydropower generation offered by the Congo river and its massive Inga falls and related river systems, and accordingly the DRC holds the highest hydroelectric potential of around 10,000 MW in Africa.[15]

Despite its richness in natural resources and economic success, the country still faces energy access challenges as only around 11% of the 75 million inhabitants have access to electricity. Reliance on biomass in the form of wood fuel is prevalent in rural areas, consequently posing environmental challenges.[16]

The electricity sector is dominated by the *Société national d'électricité* (SNE), a vertically integrated utility company. Massive investments are needed in the electricity sector, and it was this that led to significant regulatory reform in 2009, including the Electricity Code policy project, which enabled the liberalisation of the electricity sector and the creation of the Electricity Regulation Authority (ARE), the National Electrification Fund (FONEL), and the National Electrification Agency (AGENA).

Furthermore, Investment Code 2002 regulates the sector. Private investors are assisted by the National Agency for Investment Promotion (ANAPI) to facilitate investment process. The DRC is also part of the code of Organization for the Harmonization of Business Laws in Africa (OHADA), an important instrument to promote investment.

3.1.4 Kingdom of Lesotho

Landlocked Lesotho is a constitutional monarchy, with GDP projected at 2.6% which is mostly driven by mining production, construction, and government services.[17]

---

[15] The Southern African Communities' Environmental Institute: An overview of the Energy Picture for SADC Countries, with a focus on renewable energy, October 2014.

[16] Ibid.

[17] World Bank: Lesotho Overview, http://www.worldbank.org/en/country/lesotho/overview, last updated on 7 October 2015, last visited on 3 February 2016.

Access to electricity is still a major challenge, as it is estimated that only 35% of the 1.9 million population had access to electricity in 2015, although this has risen from 5% estimated to have access in 2000.[18] There is heavy reliance on biomass especially in the rural areas where the majority of people have no access to electricity.[19]

Massive energy investments are essential in order to increase access to electricity. Accordingly, the Lesotho Electricity Authority (LEA) was set up through the Lesotho Electricity Authority Act of 2002 to regulate the electricity supply industry in the country and also to promote competition. Currently there is no foreign investment law; however, there are some legislations that promote investment. For instance, the Land Act of 2010 allowed foreign investors to hold land titles as long as 20% of their company was owned by local investors. Furthermore, the time for starting up a business was reduced from forty days to five days as stipulated in the Companies Act of 2011. This Act also strengthened investor protections by increasing the disclosure requirements for related-party transactions and improving the liability regime for company directors in cases of abuse of power in related-party transactions.

Due to these policy reforms, the ranking of Lesotho improved from 153 in 2012 to 128 in 2015, according to the World Bank's Doing Business Report.[20] Lesotho is also a member of many multilateral organisations and a signatory to different trade agreements, including among others COMESA, WTO, MIGA, and ACP-EU. Lesotho's membership in SACU and the SADC market has resulted in the country having a gateway to larger regional markets.

3.1.5 Republic of Madagascar

Madagascar is an island country in the Indian Ocean, off the coast of Southeast Africa.[21] As of 2014, the country's GDP was estimated at USD

---

[18] Lesotho, Ministry of Energy, Meteorology and Water Affairs; *Expression of interest to participate in SREP*. Page 5.

[19] Ibid.

[20] U.S. Department of State: Lesotho Investment Climate Statement, June 2015, http://www.state.gov/documents/organisation/241842.pdf, last visited on 5 February 2015.

[21] The nation comprises the Island of Madagascar, the fourth largest in the world, as well as numerous smaller peripheral islands.

10.5 billion with a growth of 3.0%, mainly driven from sectors such as mining, agribusiness, banking, farming, and fishing among others.[22] Madagascar is rich in energy resources such as oil and gas. For example, the Tsimiroro is a large oil field in the onshore Morondava Basin operated by Madagascar Oil and it is estimated to contain about 1.7 billion barrels of recoverable oil resources.[23] In addition, the offshore Morondava basin is believed to have 10.8 billion barrels of oil, 167 Tcf of natural gas, and 5 billion barrels of natural gas liquids.[24]

Despite the energy resources in the country access to modern energy is still low, with only 19% of the 20.7 million inhabitants having access to electricity. It is estimated that over 90% of the population, especially in rural areas, are still reliant on traditional biomass, which is often unsafe due to the indoor air pollution it causes.[25] The national utility JIRAMA manages the national transmission system and there are three grids in the country, the Antananarivo Grid, the Toamasina Grid, and the Fianarantsoa Grid. Hydropower potential is estimated at 7800 MW but only 165 MW is exploited.[26]

The country is committed to mobilising investment in the energy sector, as well as promoting public private partnerships. Madagascar's Investment Law (2007-036) includes provisions protecting investors against the risk of expropriation and protection of patent rights. The Law further guarantees investors the freedom to transfer funds abroad without prior authorisation and it also provides for a stability clause guaranteeing investor privileges from future legal or regulatory measures.[27] Madagascar is also a member of many multilateral organisations and a signatory to different trade agreements, including among others COMESA, SADC, WTO, MIGA, and ACP-EU.

---

[22] http://www.worldbank.org/en/country/madagascar

[23] See details, EIA: *Emerging East Africa Energy*, last updated on 23 May 2013.

[24] See details, EIA: *Emerging East Africa Energy*, last updated on 23 May 2013.

[25] Ministry of Energy—Expression of Interest to participate in the Scaling Up Renewable Energy in Low Income Countries Program (SREP). Page 3.

[26] Ministry of Energy—Expression of Interest to participate in the Scaling Up Renewable Energy in Low Income Countries Program (SREP). Page 3.

[27] Madagascar Investment Climate Statement. Page 5.

3.1.6 Republic of Malawi

Malawi, a landlocked country in southeast Africa, has an economy based heavily on agriculture, with around 80% of the population living in rural areas.

Malawi is one of the least electrified countries globally. Currently only 9% of the 14.4 million people have access to electricity, and these are mostly in urban areas, as only about 1% of the rural population have access to electricity.[28] Electricity generation mostly comes from the four hydroelectric power stations on the Shire River. The irregular water flow, especially in the dry seasons, often makes power supplies unreliable.

Access to modern energy is crucial to Malawi development. The Malawi Vision 2020 advocates the provision of efficient energy supplies as a key prerequisite to achieving economic development and middle-income status.[29]

Malawi is keen to promote and protect investments in the country. The Malawi Constitution protects investment irrespective of ownership and it also prohibits expropriation without due compensation. An Investment and Export Promotion Act was also passed by the government in 2012, providing tax benefits to investors and pioneer industries in manufacturing, agriculture, and mining among others. The Malawi Investment and Trade Centre Limited, which was incorporated on 2 December 2010, also promotes and facilitates both domestic and foreign investment. Malawi is also a member of many multilateral organisations and is a signatory to different trade agreements, including among others COMESA, SADC, WTO, NEPAD, ACP-EU, MIGA, and ICSID.

3.1.7 Republic of Mauritius

Mauritius is an island nation in the Indian Ocean, with one of the most competitive and diversified economies in Africa. As of 2014, the country's

[28] The Government of Malawi, Ministry of Energy (April 2014): *Expression of interest to participate in Scaling up Renewable Energy Program in low income countries (SREP)*. Page 3.

[29] Malawi Growth and Development Strategy II (MGDS II) January 2012. Page 77.

GDP was estimated at USD 12.62 billion and it is considered an upper middle-income country.[30]

The country has no known coal, oil, or natural gas reserves and as such depends heavily on imported petroleum to meet most of its energy requirements. The country's economy is based on among others, tourism, textiles, sugar, and financial services.

Mauritius has accomplished total electrification, as 100% of the population has access to electricity. Biomass, hydro, solar, and wind are the local renewable sources.[31] Fuel wood and charcoal are minimally used in the country, and in this regard bagasse is the type of biomass mainly used contributing around 22% of the primary energy supply. It is a by-product of the sugar industry.

The legislative framework for the electricity sector is comprised of the Electricity Act of 1939 (amended 1991), the Electricity Regulations of 1939, and the Central Electricity Board Act (1964). Established under the Act, the Central Electricity Board (CEB) is the one body responsible for the control and development of electricity supply in Mauritius. It is also the supplier and generator of electricity and reports to the Ministry of Renewable Energy and Public Utilities.

The Investment Promotion Act 2000 governs investment in Mauritius. There is also a government agency whose aim is to promote and facilitate investment in the country. The Mauritius Board of Investment (MBOI) acts as a one-stop agency for business registration and guides investors through the necessary processes for doing business in the country. Mauritius is a member of several multilateral and bilateral agreements including among others ACP-EU, WTO, COMESA, the Indian Ocean Rim Association for Regional Cooperation, and the Indian Ocean Commission.

3.1.8 Republic of Mozambique

Mozambique is a resource endowed country with large untapped natural resources reserves and energy potential, including hydropower, coal, and

[30] World Bank Data, http://data.worldbank.org/country/mauritius

[31] Ministry of Energy and Public Utilities-Mauritius, retrieved from http://publicutilities.govmu.org/English/Pages/Publications.aspx, last visited on 20 February 2016.

natural gas among others. Mozambique is considered a low-income country with a GDP estimated at USD 16.3 billion in 2014.

Mozambique has proven reserves of natural gas in the Rovuma Basin Offshore estimated at 127 Tcf, which will enable LNG plants. The LNG project is estimated to generate significant fiscal revenues of around USD 115 billion over 2020–2040, which will enable development of other projects.[32] In 2014, the government approved the Mozambique Gas Master Plan, a comprehensive plan to use the gas discoveries for national development.[33]

Despite the country's massive energy resources, it is estimated that only 12% of the 23 million population have access to electricity in Mozambique. More investments are needed in the electricity sector, such as Law No. 21/97, liberalised the sector allowing private investors to set up power generation projects. Private investors can apply for a concession to generate electricity and supply the utility company *Electricidade de Moçambique* (EDM).[34] Furthermore, Mozambique's Law on Investment, No. 3/93, offers incentives and guarantees to national and foreign investors to protect and promote investments. Among others, the law protects property rights and intellectual property rights, prohibits expropriation and nationalisation unless for public purpose, and payment of compensation and transfer of funds, including fiscal and custom incentives such as exemption from taxes, levies, import duties, international arbitration.[35]

Moreover, Mozambique has approved PPP Law No. 15/2011 dated August 10, which sets out rules and procedures to promote large infrastructure projects as well as national and foreign private sector participation. The government's Investment Promotion Center (CPI) is the institution responsible to coordinate and facilitate investments. The country is also a member of several multilateral agreements including among others the Overseas Price Investment Corporation (OPIC), WTO, and ACP-EU.

---

[32] IEA Energy Outlook 2014. Page 156.

[33] Mozambique Gas Master Plan. Page.

[34] Mozambique Law No. 21/97 concerning production, transport, supply, and sale of electric power and creating the National Council for Electricity.

[35] See, articles 13, 14, 15, 16 of the Law n. 3/93, of 24 June.

3.1.9 Republic of Namibia

Namibia is one of the least densely populated countries in the world, due in large part to the presence of the Namib Desert. Namibia's GDP was estimated at USD 13.4 billion in 2014 and the country is considered an upper middle-income country. Agriculture, tourism, and herding form the basis of Namibia's economy. The country is also rich in resources such as diamonds, uranium, copper, gold, and silver among others, and as such mining contributes highly to the country's economy.

Namibia discovered natural gas in offshore fields representing an opportunity to develop downstream activities. Currently, refined petroleum products are all imported from South African refineries and other sources in West Africa and the Middle East.[36]

Despite its massive natural resources, energy access is still a challenge as it is estimated that only 34% of the 2.1 million people have access to electricity and these are mostly in the urban areas. As such, the majority of the people especially in rural areas rely on biomass in the form of wood fuel and charcoal, which has enormous negative effects on both the environment and health of the people.

Namibia Power Corporation (Proprietary) Limited (NamPower) dominates the power sector in the country. NamPower is a vertically integrated electricity utility with a monopoly over generation and transmission.[37] The Electricity Control Board (ECB) regulates the sector and exercises control over the electricity supply industry, and it has the responsibility for promoting private sector investment in the electricity industry. The Electricity Act 2007 is the principal legislation in the electricity sector. The Act established the regulatory framework for the power sector implemented by the ECB.[38]

More investment in the electricity sector is essential to reach universal energy access in Namibia.

[36] Energy Policy scenarios for Namibia from the present time to 2050. Page 56.

[37] NamPower's origins are as a private company established by South Africa to operate the Ruacana hydropower project in 1964.

[38] The Electricity Act 2007 replaced the Electricity Act of 2000. The recent Act is proposed to be superseded by an Electricity Bill and an Energy Regulator Bill.

The government is committed to attracting foreign investors in the country. Foreign Direct Investment (FDI) is governed by the Foreign Investment Act of 1900, which guarantees equal treatment for foreign investors and Namibian firms, including among others the right to remit profits and access to foreign exchange, fair compensation in the event of expropriation, and international arbitration of disputes between investors and the government. The Namibia Investment Center (NIC) was also established under the Act to serve as the country's official investment promotion and facilitation office. The country is also a member of several multilateral agreements including among others WTO, ACP-EU, and MIGA.

3.1.10 Republic of Seychelles

Seychelles is an archipelago of 115 islands in the Indian Ocean off East Africa. It is the least populous country in SADC with a population of around 87,000. Seychelles' GDP was estimated at USD 1.406 billion in 2014 and it is considered a high-income country.[39]

Seychelles' economy is heavily dependent on among others fishery, tourism, boat building, processing of coconuts, and vanilla. Seychelles has a very high level of electrification, estimated at 96%. However, the country still faces challenges in developing the renewable energy sector. The country is highly dependent on diesel imports for power and this has resulted in high prices, low energy security, and constraints on growth and development. Based on this, the government set a renewable energy target of 15% by 2030. This would require significant investment in all forms of renewable energy, but primarily in wind, solar, waste-to-energy, and bio fuels.[40]

The Public Utilities Corporation (PUC) is a government owned corporate body responsible for providing electricity, water, and sewerage services.[41] The PUC generates approximately 95% of the country's electricity, with the remaining 5% being self-generators for own use.[42] With regard

---

[39] See World Bank Data, http://data.worldbank.org/country/seychelles

[40] Government of Seychelles 2012, Rio +20, National Report. Page 46.

[41] PUC was established under the Public Utilities Corporation Act, Cap 196.

[42] See, PUC website http://www.puc.sc/, last visited on 29 January 2010, also http://www.reegle.info/policy-and-regulatory-overviews/SC, last visited on 2 October 2015.

to foreign investment promotion and protection, in 2010 the government enacted the Foreign Investment Act, which sets out a legal framework that is conducive to greater flows of investments. It provides fair and equitable treatment, protection of property rights against expropriation, ensures prompt and adequate compensation in the case of expropriation, transfer of capital and profits, and framework to PPP.[43] The Seychelles Investment Board (SIB) serves as the investment promotion agency for the country and advises all foreign investors on the laws, regulations, and procedures for their activities in Seychelles.[44] The country is also a member of several multilateral agreements including among others WTO, ACP-EU, and MIGA.

3.1.11 Republic of South Africa

South Africa (SA) is the continent's second largest economy with a GDP of USD 349.8 billion and is considered an upper middle-income country. It is very rich in natural resources such as chromite ore, diamonds, gold, platinum, and manganese among others. The mining industry contributes significantly to the country's economy.

The country is also the biggest energy producer in SADC and is endowed with energy resources such as coal and natural gas, and renewable energy resources including solar. It is estimated that around 77% of SA energy needs are directly derived from coal, with reserves estimated at around 33 billion tons, accounting for 95% of total African and 3% of the world's reserves.[45]

The country has achieved total electrification. However, the country is highly dependent on coal leading to a challenge in switching to cleaner energy. As a result, SA has set a target of 33% of total electricity to come from renewable sources in 2030.[46]

---

[43] The Foreign Investment Act 2010, available at http://www.wipo.int/edocs/lexdocs/laws/en/sc/sc011en.pdf

[44] Seychelles Investment Climate Statement. Page 4.

[45] USA EIA. Page 3.

[46] According to the Integrated Resource Planning, 2010 (updated in 2013).

The government owned national power utility Eskom dominates the country's electricity sector, with 27 operational power plants generating over 95% of the country's electricity and over 40% of all electricity on the African continent making it one of the ten largest power utilities in the world.[47]

More investments, especially in the renewables sector, are needed in the future to ensure the transition to cleaner energy. Trade and Investment South Africa (TISA), an investment and promotion agency, aims to increase export capacity and support direct investment flows. TISA's strategic goals include to increase the quality and quantity of foreign and domestic investment; and undertake effective investment recruitment campaigns.[48] The SA economy is also advanced, offering investors protection of property rights and making expropriation illegal except in certain circumstances such as in the public interest, and the paying of prompt and adequate compensation. The country is also signatory to bilateral and multilateral agreements such as WTO and SACU among others.

3.1.12 The Kingdom of Swaziland

Swaziland, officially known as the Kingdom of Swaziland, is a landlocked country with a GDP of GDP 3.400 billion and is considered a lower middle-income country. The country is rich in natural resources such as asbestos, small gold and diamond deposits, and quarry stone among others. The country has no known oil or natural gas reserves, although it is rich in coal and renewable resources such as solar. Coal reserves are estimated at 158.73 million short tons.

Access to electricity is still a major challenge in the country. It is estimated that 55% of the 1 million inhabitants have access to electricity, mostly in urban areas. As such, the majority of people in rural areas rely on traditional biomass in the form of wood fuel and charcoal.[49]

---

[47] USA EIA. Page 9.

[48] TISA is a subdivision of the Department of Trade and Industry.

[49] IRENA—International Renewable Agency—Swaziland Renewables Readiness Assessment. Page 7.

Swaziland Electricity Company (SEC) is engaged in the business of generation, transmission, and distribution of electricity and the import and export of electricity within and outside Swaziland.[50] The Electricity Act of 2007 governs the electricity sector and provides for the regulation of the electricity supply industry. The Act regulates the generation, transmission, distribution, and supply of electricity in the country.[51] The Energy Regulatory Authority (ERA) was also established under the ERA Act of 2007, with powers and functions including receiving and processing applications for licenses; modifying/varying licences; and monitoring the performance and the efficiency of licenced operators among others.

Investments are needed to further develop the country's energy sector. The Swaziland Investment Promotion Act regulates the investment sector and offers protection to investors and their property. The Act stipulates that investors' property shall not be compulsorily acquired except in certain circumstances including in accordance with applicable legal procedures; in public interest; done without discrimination and upon prompt payment of adequate and fair compensation.[52] The Act also established the Swaziland Investment Promotion Authority (SIPA) to promote and facilitate foreign direct and local investment in the country, and it provides a one-stop information and support facility to investors.[53] There are also several incentives for investors including the repatriation of profits and fully serviced industrial sites. Financial incentives for all investors also include tax allowances and deductions for new enterprises, exemption from withholding tax on dividends, and a low corporate tax.

3.1.13 United Republic of Tanzania

Tanzania, officially the United Republic of Tanzania, is located in East Africa and has a GDP of around USD 49.1 billion. The country is rich

[50] The Swaziland Electricity Company Act of 2007 transformed Swaziland Electricity Board into a company called Swaziland Electricity Company.

[51] The Electricity Act of 1963 was repealed as a result of the promulgation of the Electricity Act of 2007 on 1 March 2007.

[52] Section 20 of the Swaziland Investment Promotion Act, 1998.

[53] SIPA is established under Section 3 of the Swaziland Investment Promotion Act, 1998.

in mineral resources such as coal, gold, diamond, nickel, and uranium among others.

Tanzania also has a variety of energy resources, which include hydropower, natural gas, coal, biomass, geothermal, solar, and wind power. The nation has proven natural gas reserves estimated at 53.28 TCF, composed of deep sea (45.28 TCF) and onshore reserves (8 TCF). Tanzania has approved the Natural Gas Policy 2013, a framework to promote national development based on the natural gas industry.[54]

Despite the available energy resources, access to electricity is still low in the country with only 24% of the 44.5 million inhabitants having access to electricity, and a low 7% electrification rate in rural areas. Therefore, there is high reliance on biomass especially in the rural areas.[55] About 88% of total energy consumption is biomass in the form of firewood and charcoal for cooking. Commercial energy sources such as oil, gas, electricity, and coal, as well as non-biomass renewable energy, account for the remaining 12%.[56]

The Electricity Act of 2008 governs the electricity sector in Tanzania. The Act liberalised the production and distribution of electricity, which was previously reserved for the Tanzania Electric Supply Company (TANESCO). The Act also provides for the facilitation and regulation of the generation, transmission, transformation, distribution, supply, and use of electric energy, cross-border trade in electricity, and the planning and regulation of rural electricity in Tanzania.[57]

The Government of Tanzania has been engaged in attracting foreign investment.[58] The Tanzanian Investment Act of 1997 (TIA) is the main legislation governing investment in the country. The Tanzania Investment Center (TIC) operates as a one-stop agency to coordinate, encourage,

---

[54] See Energy Outlook. Page 157. Further, Tanzania Natural Gas Policy 2013. Page 9, available at http://www.tanzania.go.tz/egov_uploads/documents/Natural_Gas_Policy_-_Approved_sw.pdf, last viewed on 10 September 2015.

[55] Ministry of Minerals and Energy (2014), Electricity Supply Reform Strategy and Roadmap, 2014–2025 June 2014. Page 2.

[56] Energy Charter secretariat; *Tanzanian energy sector under the universal principles of the Energy Charter*. Page 6.

[57] The Electricity Act, 2008, Laws of Tanzania.

[58] Tanzania Investment Climate Statement. Page 4.

promote, and facilitate investment in Tanzania.[59] TIC promotes investment and trade opportunities in agriculture, mining, tourism, telecommunications, financial services, energy, and transportation infrastructure.[60]

3.1.14 Republic of Zambia

Zambia is a landlocked country, with a GDP estimated at USD 27 billion and it is considered a lower middle-income country. The country is endowed with natural resources such as copper, gold, nickel, and iron among others.

The country's energy resources include biomass, electricity, coal, petroleum, and renewable energy. Zambia is rich in coal resources, which are estimated to be about 80 million tonnes, located in the Southern Province, with other probable coal reserves in Luangwa North, Luano, and Lukusashi among others.[61]

The country also has huge hydro resources and is estimated to possess 40% of the water resources in SADC, although Zambia is estimated to have developed only 2177 MW. Around 6000 MW of hydro potential is still unexploited. Petroleum is wholly imported in the country.[62]

Despite this richness in energy resources, only around 22.1%[63] of the 13.5 million people in Zambia have access to electricity and these are mostly based in rural areas, where it is estimated that 22% are electrified compared to 4.5% in urban areas. This differs from most African countries, where it is usually the urban areas which are highly electrified. There is high reliance on biomass in the form of wood fuel, which has negative effects such as indoor air pollution. Thus, more investment is crucial in the energy sector to overcome this challenge.

---

[59] See, article 5, Tanzania Investment Act 1997, and Tanzania Investment Centre, at http://www.tic.co.tz/

[60] Tanzania Investment Climate Statement. Page 4.

[61] Zambia Development Agency: Zambia Energy Sector Profile, September, 2014.

[62] Zambia Development Agency: Zambia Energy Sector Profile, September, 2014.

[63] The World Bank: Access to electricity data, http://data.worldbank.org/indicator/EG.ELC.ACCS.ZS?order=wbapi_data_value_2012+wbapi_data_value+wbapi_data_value-first&sort=asc, last visited on 20 February 2016.

The Energy Regulation Act 1995 governs the energy sector. The Act repealed the National Energy Council Act and the Zambia Electricity Supply Act, and established the Energy Regulation Board (ERB) with the following functions: issuing licenses, and developing and implementing appropriate rules to promote competition in the energy sector.[64] Zambia Electricity Supply Company (ZESCO) is the state-owned electricity utility company. ZESCO is a vertically integrated utility that generates, transmits, distributes, and supplies electricity. It is the largest power company producing about 80% of the total electricity consumed in the country.[65]

The country is keen to attract investors in the energy sector. Zambia Development Agency (ZDA) is the government agency responsible for engaging with foreign investors. Established under the ZDA Act 2006, it aims to operate as a one-stop facility to coordinate and facilitate investment in the country. The ZDA Act offers investors promotion and protection incentives. For example, property can only be expropriated for public purposes, with the consequent right to demand fair market value compensation.[66]

3.1.15 Republic of Zimbabwe

Zimbabwe is a landlocked country located in southern Africa, between the Zambezi and Limpopo Rivers. The country's GDP was estimated at USD 13.6 billion in 2014.[67]

The country's economy is highly dependent on agriculture, mining, manufacturing, and tourism.[68] Mineral resources include copper, iron ore, diamonds, and tin among others.

The country has no oil or gas resources and the country depends on imports. However, coal resources are estimated at 30 billion tons. With

[64] Energy Regulation Act, Chapter 436.

[65] Other electricity companies include: Copperbelt Energy Corporation and the Lusemfwa Electricity Company.

[66] Zambia Development Agency Act No. 6 of 2006.

[67] World Bank: Zimbabwe country profile, http://www.worldbank.org/en/country/zimbabwe

[68] Industrial Development Policy (IDP) 2012–2016. Page v.

regard to electricity, only 40.5% of the 12.8 million inhabitants have access.[69] Zimbabwe Electricity Supply Authority (ZESA) is a state-owned company whose task is to generate, transmit, and distribute electricity in Zimbabwe.[70]

The Electricity Act of 2002 is the main legislation governing the sector, leading to the liberalisation of the electricity market and allowing private companies to acquire licenses to generate and supply electricity.[71] The Act also restructured and unbundled ZESA from a vertically integrated utility into separate successor companies. The companies include Zimbabwe Power Company (ZPC), Zimbabwe Electricity Transmission Company (ZETCO), and Zimbabwe Electricity Distribution Company (ZEDC).[72]

Massive investments are needed in the electricity sector. The Constitution guarantees the right to private property and it prohibits expropriation of investor property without adequate compensation.[73] The Zimbabwe Investment Authority is the Government agency responsible to secure investments.[74] Zimbabwe also has a number of incentives designed to attract FDI including tax breaks for new investment by foreign and domestic companies, and allowing capital expenditures on new factories, machinery, and improvements to be fully tax deductible. The government also waives import taxes and surtaxes on capital equipment. The country is also a signatory to several multilateral treaties including MIGA, Overseas Private Investment Corporation (OPIC), International Convention on Settlement of Investment Disputes, and the New York Convention on the enforcement of foreign Arbitral Awards.

---

[69] The World Bank: Access to electricity data, http://data.worldbank.org/indicator/EG.ELC.ACCS.ZS?order=wbapi_data_value_2012+wbapi_data_value+wbapi_data_value-first&sort=asc, last visited on 20 February 2016.

[70] See ZESA website http://www.zesa.co.zw/, last visited on 29 January 2016.

[71] See article 42 of the Electricity Act 2002.

[72] Article 68 of the Electricity Act 2002.

[73] Section 71 (2) of The Zimbabwe Constitution, 2013.

[74] The Authority was established under Section 3 of the Zimbabwe Investment Authority Act (Chapter 14:30).

3.2 The Concept of Regionalism in the SSA Energy Sector

3.2.1 Panoramic Overview of Regionalism

Regionalism is defined as the expression of a common sense of identity and purpose combined with the creation and implementation of institutions that express a particular identity and shape collective action within a geographical region.[75] In this book, regionalism is defined in the context of international relations and is discussed as one of the three constituents of the international commercial system—the other two being multilateralism and unilateralism. This definition brings into play the component of geography, and mostly connotes the idea of an international region, which is understood to mean a limited number of states linked by a geographical relationship and by a degree of mutual inter-independence: this can also be called international regionalism, which is understood as the formation of interstate associations or groupings on the basis of region.[76]

Though the above definition focuses on the issue of geography and territory, it is imperative to note that there are some regional blocks which do not purely consider geography as the basis for inclusion of new members. The case in point is the League of Arab States (LAS) which essentially applies linguistic-cultural aspects as the basis for the inclusion of members rather than geographical conditions.[77] The Organisation of Islamic Cooperation (IOC) also focuses on religious identity in defining its membership rather than geography.[78] Furthermore, other regional

[75] Ethier, W., 1998. *The international commercial system (No. 210)* (p. 11). Princeton University International Economics.

[76] Nye, J.S. ed., 1968. *International regionalism: Readings*. Little, Brown.

[77] The Arab League is a regional organisation of Arab countries in and around North Africa, the Horn of Africa, and Arabia. It was established in Cairo on 22 March 1945 with six members; to date, it includes twenty-two member states.

[78] The IOC is the collective voice of the Muslim world, which was established upon a decision of the historical summit which took place in Rabat, Kingdom of Morocco on 12th Rajab 1389 Hijra (25 September 1969) following the criminal arson of Al-Aqsa Mosque in occupied Jerusalem. The OIC is the second largest inter-governmental organisation after the United Nations with a membership of 57 states spread over four continents.

blocs are basically driven by a specific function and purpose.[79] A case in point is the North American Free Trade Agreement (NAFTA)[80] and the Asia-Pacific Economic Cooperation (APEC)[81] which are mostly focused on economic integration. On the other hand, the creation of some regional blocs was initially driven by security concerns as is the case with the North Atlantic Treaty Organization (NATO)[82] and the Shanghai Cooperation Organization (SCO).[83]

In an attempt to link the concepts of regionalism and region-building, Louise Fawcett describes regionalism as a looser term that is understood as the promotion of regionally based policies and practices. Fawcett is conscious to distinguish both regionalism and regional building from the concept of regionalisation when he argues that the latter can refer to spontaneous or undirected regional activity.[84] Ravenhill also supports the idea of distinguishing regionalism from regionalisation, and as such he refers to the latter as a process of increased integration within a given geographical area.[85]

Historically, the notion of regionalism was influenced by various changes in the international system including the Cold War and its ending, the post-hegemonic era, and the processes of decolonisation and globalisation.[86] Regionalism dates back as far as the 1950s and 1960s with

---

[79] Fawcett, L., 2016. Region-building debates in a global context. In *Region-building in Africa* (pp. 21–36). Palgrave Macmillan US.

[80] NAFTA is an agreement signed by Canada, Mexico, and the United States, creating a trilateral trade bloc in North America.

[81] APEC was established in 1989 in response to the growing interdependence of Asia-Pacific economies, and it currently has twenty-one members.

[82] NATO is an inter-governmental military alliance consisting of twenty-nine member countries across North America and Europe.

[83] SCO is a Eurasian political, economic, and security organisation with eight member states including China, Kazakhstan, Kyrgyzstan, Russia, Tajikistan, Uzbekistan, India, and Pakistan. Though SCO was officially created in 2001, its origin dates as far back as the 1990s, with the signing of the Treaty on Deepening Military Trust in Border Regions in Shanghai, China by the heads of States of China, Kazakhstan, Kyrgyzstan, Russia, and Tajikistan.

[84] Fawcett, L., 2016. Region-building debates in a global context. In *Region-building in Africa* (pp. 21–36). Palgrave Macmillan US.

[85] Ravenhill, J., 2016. Regional integration in Africa: Theory and practice. In *Region-building in Africa* (pp. 37–52). Palgrave Macmillan US. Page 36.

[86] In distant centuries however, regionalism was influenced by powerful states and empires that sought to build and develop regions as an extension or demonstration of their power or as a means

the establishment of Western Europe—this is understood as the old regionalism, which was mostly influenced by the Cold War logic. The Cold War is indeed seen as being essential in providing an arena for the selective development of regional organisations and as a foundation for later regional building.[87]

In the 1980s, the global community experienced another form of regionalism typically referred to as "new regionalism". The post-hegemonic condition of the post–Cold War international system has indeed been cited as one of the reasons for the new regionalism. Historically, regional blocs and regionalism were considered as better alternatives after the breakdown of the world order in the 1930s, which was due to the collapse of the British hegemony. The world order which was sustained by the British hegemony in the nineteenth century had fallen apart in the twentieth.[88] Gamble, citing E.H. Carr, describes the world order as "the golden age of continuously expanding territories and markets of a world policed by the self-assured and not too onerous British hegemony, of a coherent 'western civilization' whose conflicts could be harmonized by a progressive extension of the area of common development and exploitation".[89] After the Second World War, the United States emerged as the new hegemony; however, the country was also unable to continue to play the role of hegemony and to supply the public goods necessary to stabilise the global economy, and at this time many observers agreed that the future now belonged to national economies and regional blocs.[90]

New regionalism has also been associated with the UN and there is a general consensus that the success story of the EU integration has

---

of balancing the power of others. For a full discussion on this, see, Fawcett, L., 2016. Region-building debates in a global context. In *Region-building in Africa* (pp. 21–36). Palgrave Macmillan US.

[87] Fawcett, L., 2016. Region-building debates in a global context. In *Region-building in Africa* (pp. 21–36). Palgrave Macmillan US.

[88] Padoan, P.C., 2001. Political economy of new regionalism and world governance. In MarioTelò, *European Union and new regionalism: Regional actors and global governance in a post-hegemonic era* (pp. 39–59) Burlington: Ashgate, at 25.

[89] Ibid.

[90] Padoan, P.C., 2001. Political economy of new regionalism and world governance. In MarioTelò, *European Union and new regionalism: Regional actors and global governance in a post-hegemonic era* (pp. 39–59) Burlington: Ashgate, at 25.

provided both the inspiration and normative model for the new wave of regionalism throughout the world.[91] Regional blocs including EAC and ECOWAS have been trying to imitate the EU model, and in some instances the EU policies in these regions have acted as an external driver to regionalism. Besides the influence of the EU, other external drivers have been felt from international organisations such as the UN, World Bank, and multilateral institutions such as the General Agreement on Tariffs and Trade (GATT) and World Trade Organisation (WTO).[92] Besides external drivers, there are also internal drivers which of course reflect the specific concerns of the states involved, unlike the external drivers which mostly focus on the interests of international organisations.

One would indeed wonder why regionalism started in Europe and not in other parts of the globe. This issue has been addressed by some scholars and there are several reasons cited including that the conditions in many newly independent states were not conducive for regional integration, due to inexperience on the part of leaders and institutions; the actions of superpowers placed restraints on regional autonomy as a way of strengthening their regionalism in other forms such as NATO; some international liberals were against regionalism due to the fear that it would threaten aspirations for global peace.[93]

In recent years, the concept of regionalism has attracted efforts to distinguish between "old" and "new" regionalism and this has gained academic attention over the last two decades or so.[94] Table 3.1 sets out a

---

[91] Telò, M. ed., 2013. *European Union and new regionalism: Regional actors and global governance in a post-hegemonic era*. Ashgate Publishing, Ltd.

[92] Fawcett, L., 2016. Region-building debates in a global context. In *Region-building in Africa* (pp. 21–36). Palgrave Macmillan US.

[93] Fawcett, L., 2016. Region-building debates in a global context. In *Region-building in Africa* (pp. 21–36). Palgrave Macmillan US.

[94] For instance, in Keating, M., 1998. *The new regionalism in Western Europe: Territorial restructuring and political change*. Cheltenham, UK: E. Elgar; Lovering, J., 1999. Theory led by policy: The inadequacies of the 'new regionalism' (illustrated from the case of Wales). *International Journal of Urban and Regional Research, 23*(2), pp. 379–395; Hettne, B. and Söderbaum, F., 2006. Regional cooperation: A tool for addressing regional and global challenges. In *Meeting global challenges: International cooperation in the national interest. Final report* (pp. 179–244); and recently in Hettne, B., Sunkel, O. and Inotai, A. eds., 2016. *Globalism and the new regionalism* (Vol. 1). Springer.

3.2.2 Origin of Regionalism in Sub-Saharan Africa

Although the European form of regionalism has had an impact on the current trends in the African continent, it is worth noting that the desire for regionalism and advocacy for togetherness and oneness of all Africans have been in existence long before the success stories of the EU regionalism. Regionalism in Africa has always been based on the desire to unite a continent which was divided by former colonialists.

Historically, regionalism in Africa and the desire for regional cooperation have been evident since the independence of most African states and have their roots in political Pan-Africanism, which required a lot of solidarity to achieve the various goals of the liberalisation movements. The fundamental purpose of Pan-Africanism was national independence leading to African unity.[99]

In his book *Africa must unite*, Kwame Nkurumah, Ghana's first president, was very optimistic of a united Africa. Although he acknowledged that there were some obstacles to uniting Africa such as differences in culture, race, and language, he nevertheless believed that the forces making for unity far outweighed those which divided Africa.[100] Moving towards a unified Africa, five Pan-African Congresses were held, the first one taking place in Paris in 1919 and the fifth one taking place in Manchester in 1945.[101] Kwame Nkrumah's views and ideologies were in support of a federation of all African countries and these were indeed supported by other countries under the umbrella of the Casablanca block.[102] Besides the Casablanca group, there also existed the Monrovian

[99] Nkrumah, K., Arrigoni, R. and Napolitano, G., 1963. *Africa must unite* (p. 97). London: Heinemann. P. 133

[100] Nkrumah, K., Arrigoni, R. and Napolitano, G., 1963. *Africa must unite* (p. 97). London: Heinemann.

[101] The second Pan-African Congress was held in London in 1921: two years later the third congress was held in London in 1923 and among the resolutions passed in the third congress was one asking the voice for Africans in their own governments and one asking for the right of access to land and its resources. The fourth Pan-African Congress was held in New York in 1927 and the fifth one was held in Manchester in 1945. For details see, Nkrumah, K., Arrigoni, R. and Napolitano, G., 1963. *Africa must unite* (p. 97). London: Heinemann. P. 133.

[102] The Casablanca block was led by Kwame Nkrumah of Ghana and it was supported by other African countries including Algeria, Guinea, Egypt, Morocco, Mali and Libya.

bloc, which basically was against the notion of political federation.[103] In an effort to dissolve the disputes among these two groups a meeting was called in Ethiopia and as a result the Organisation of African Unity (OAU) was established on 25 May 1963 in Addis Ababa, with the main objectives of coordinating and intensifying the cooperation of African States and defending the sovereignty, territorial integrity, and independence of African States just to name but a few. Moreover, it also actively supported movements against all forms of colonialism and white minority rule in Africa.[104] The OAU was later dismantled on 9 July 2002 and replaced with the African Union (AU).[105]

It is generally agreed that the need to solve conflicts related to colonialism, anti-colonialism, apartheid, wars of liberation, and the Cold War was the basis for forming regional organisations in Africa and this is clearly manifested in the current SADC.[106] The creation of SADC had its origins in the struggle to liberate white minority rule in Rhodesia (now Zimbabwe), South Africa, and South-West Africa (now Namibia), which led to the creation of the Front Line States (FLS).[107] The FLS was later transformed into the Southern African Development Coordination Conference (SADCC) in 1980, and this was later turned into SADC.[108] The steps taken to create SADC clearly outline the main basis for regionalism and regional cooperation in Africa, which was driven by political and social desires.

Besides political and social reasons, there are also various reasons why regionalism was and is still being embraced in Africa and in this respect SSA. Economically, regional and continental integration would remove

---

[103] The Monrovvian bloc was led by Senghor of Senegal and supported by Nigeria, Liberia, and Ethiopia just to name a few.

[104] For instance, it supported the African National Congress (ANC) struggles against white minority rule in South Africa.

[105] The AU has its origin in not only the establishment of the OAU, but also the Union of African States that was established by Haile Selassie and Kwame Nkurumah in the 1960s.

[106] Omari, A.H. and Macaringue, P., 2007. Southern African security in historical perspective. *Security and Democracy in Southern Africa.*

[107] FLS originally comprised of Angola, Botswana, Mozambique, Tanzania and Zambia. The FLS created SADCC in 1980 alongside Zimbabwe, Lesotho, and Swaziland.

[108] See Omari, A.H. and Macaringue, P., 2007. Southern African security in historical perspective. *Security and democracy in Southern Africa*; Khadiagala, G.M., 1994. *Allies in adversity: The frontline states in Southern African security, 1975–1993*. Ohio University Press.

political barriers to the movement of goods, services, and factors of production. Moreover, considering that most African countries are small economies,[109] regionalism would provide the size and diversity that such economies needed to be self-sustaining.[110] The ambitious need to ensure the political and economic independence of African countries is illustrated by some ideologies such as the one advocating for the total destruction of political and economic African borders.[111] However, few Pan-African leaders agreed with total integration. Mwalimu Julius Nyerere, Tanzania's first president, noted in his speech in Zambia:

> *I believe that a real dilemma faces the Pan-Africanist. On the one hand is the fact that Pan-Africanism demands an African consciousness and an African loyalty; on the other hand, is the fact that each Pan-Africanist must also concern himself with the freedom and development of the nations of Africa.*

He therefore suggested the need for each country to promote its own nationhood to avoid internal conflict and further disunity among the African nations.[112]

From this, we can clearly determine the conflict in ideologies which made it hard for the Pan-Africanism dream of a United Africa to come to pass. These differences in ideologies were also manifested in some SSA regional organisations as far back as the 1970s. In the SSA particularly, the threat to regional economic communities as a mechanism of regional cooperation was experienced in Eastern Africa, when the EAC based on regional cooperation and founded in 1967 was dissolved in 1977,[113] due

[109] There are some big economies such as South Africa, Kenya, Nigeria, and Egypt.

[110] Levine, D.H. and Nagar, D. eds., 2016. *Region-building in Africa: Political and economic challenges*. Springer, quoting, Adedeji, A., 1935. *Comparative strategies of economic decolonization in Africa*. na.

[111] Levine, D.H. and Nagar, D. eds., 2016. *Region-building in Africa: Political and economic challenges*. Springer, referring to Amilcar Cabral, who argued that national liberation should be seen as a necessary precursor to a process of class struggle that would culminate in the destruction of capitalism and statist forms of organisation. See Cabral, A., 1979. Presuppositions and objectives of national liberation in relation to social structure. Amilcar Cabral, *Unity and struggle: Speeches and writings of Amilcar Cabral* (pp. 119–137). New York and London: Monthly Review Press.

[112] See Nyerere, J.K., 1966. The dilemma of the Pan-Africanist. *Address to the University of Zambia Congregation, 13*, pp. 207–217.

[113] The East African Community was founded by Kenya, Uganda, and the United Republic of Tanzania.

to the lack of strong political will, the disproportionate sharing of benefits of the community among the partner states, and lack of adequate policies to address the situation.[114] Although the EAC was revived in 1999 with the signing of the Treaty establishing the East African Community,[115] the question to be asked in the twenty-first century is whether this regional cooperation will be effective and if it will be maintained, given the increase in the number of member states from three to six.[116]

Drawing from the discussion here, then, we note that regional efforts to tackle energy challenges in SSA might also be threatened by not only the differences in ideologies but also the selfish desire for each country to develop its own energy sector. Additionally, the disparities among member states of the SSA regional organisations may promote selfish desires to develop their economic and energy sectors at the expense of the other member states. For instance, SADC has a population of over 298 million; however, there are social and economic disparities among SADC member states, for instance disparities in country GDP, from USD 1.3 billion in the Seychelles to USD 349 billion in South Africa. The same applies for GDP per capita, ranging from USD 342 in Malawi to USD 15,347 in the Seychelles.[117] These disparities are also present in the energy sector and thus pose a barrier to regional cooperation as a mechanism to tackle energy access.

3.2.3 Legal Basis for Regional Cooperation in SSA

As will be discussed in the next chapters, at a regional level, there is a strong desire to cooperate in addressing the challenge of energy access in SSA. However, various issues of concern arise as to the commitment and financial support together with capacity building needed to ensure and strengthen regional energy cooperation. Besides regionalism being encouraged in various regional legal frameworks, there is encouragement

[114] See the Preamble for the Treaty establishing EAC, of 1999.

[115] The Treaty for the Establishment of the East African Community, 1999.

[116] The three founding members were Kenya, Tanzania, and Uganda. Burundi and Rwanda became members in 2007 while South Sudan gained accession in April 2016.

[117] SADC Renewable Energy and Energy Efficiency Status Report, 2015. Page 9.

and support from international legal frameworks also. A case in point is Article 53 of the UN Charter, which is to the effect that "*the UN Security Council shall, where appropriate, utilize such regional arrangements or agencies for enforcement action under its authority*". Moreover, the internal structure of the UN also recognises regional efforts as envisaged in its commissions such as the UN Economic Commission for Africa (UNECA).

The relevant legal provisions which are the basis for regional cooperation in the energy sector are illustrated in the establishing regional Treaties below, and these will be further expounded on throughout the book:

- The Treaty for the Establishment of the East African Community, 1999 (EAC Treaty)

Article 2 of the EAC Treaty establishes the East African Community. The main objective is enshrined in Article 5(1) which states that:

> *The objectives of the Community shall be to develop policies and programmes aimed at widening and deepening co-operation among the Partner States in political, economic, social and cultural fields, research and technology… for their mutual benefit.*[118]

The above provision clearly illustrates the objective of deepening cooperation among the EAC member states.[119] With regard to the energy sector, Article 101 (1) states as follows:

> *The Partner States shall adopt policies and mechanisms to promote the efficient exploitation, development, joint research and utilisation of various energy resources available within the region.*[120]

Article 101 further provides the different forms the partner states can cooperate in the energy sector including promoting the least cost development and transmission of electric power; jointly plan, train, and

[118] Article 5(1) of the Treaty for the Establishment of the East African Community, 1999.

[119] Member states include Burundi, Kenya, Rwanda, Tanzania and Uganda.

[120] Article 101 (1) of the Treaty for the Establishment of the East African Community, 1999.

exchange information on the exploration, exploitation, development, and utilisation of available energy resources; development of integrated policy on rural electrification; and joint efforts in construction of oil and gas pipelines to mention but a few.[121]

- The Treaty of 28 May 1975 establishing the Economic Community of West African States (ECOWAS)

ECOWAS was established under Article 2 of the ECOWAS Treaty. The main aim of the Community is stipulated in Article 3 which states that:

> *The aims of the Community are to promote co-operation and integration, leading to the establishment of the economic union in West Africa in order to raise the living standards of its people…enhance economic stability…*[122]

In 2003, ECOWAS created an Energy Protocol with the aim of promoting further cooperation in the energy sector among the member states.[123]

- Treaty of the Southern Africa Development Community, 1992 (SADC Treaty)

SADC is established under Article 2 of the 1992 Treaty with its headquarters at Gaborone, Republic of Botswana. The objectives of SADC include achieving development, economic growth, and alleviating poverty through regional integration.[124] The Treaty encourages member states to cooperate in areas such as infrastructure, natural resources, and the environment.[125] The SADC Protocol on Energy was passed in 1996 with the main aim of ensuring regional cooperation in the energy sector.[126]

---

[121] Article 101 (2) of the Treaty for the Establishment of the East African Community, 1999.

[122] Article 3(1) of the ECOWAS Revised Treaty.

[123] ECOWAS Energy Protocol, 2003.

[124] Article 5(1) (a) of the 1992 SADC Treaty.

[125] Article 21 (3) of the 1992 SADC Treaty.

[126] SADC Protocol on Energy, 1996.

Based on the provisions above, this book will critically analyse the available forms of cooperation being employed by the EAC, ECOWAS, and SADC in the energy sector, specifically looking at energy infrastructure developments, renewable energy developments, and harmonisation of energy regulations in the region.

3.3 Why Regional Cooperation in the SSA Energy Sector?

Regional economic communities are indeed good examples of a regional cooperation mechanism. These have been established in different regions of the globe and in addition to the three mentioned already (EAC, ECOWAS, and SADC) they include the Arab Maghreb Union (AMU), The Community of Sahel-Saharan States (CEN-SAD), The Common Market for Eastern and Southern Africa (COMESA), and The European Union (EU). In this section, advantages of regional cooperation and regionalism will be discussed under two sub-sections; comparing them to isolationism and globalism, respectively.

3.3.1 Regionalism Versus Isolationism

Basically, isolationism refers to the policy employed by a country to decline from being involved in international agreements, alliances, and commitments. This policy also ensures that the country stays away from the affairs or interests of other groups (especially political affairs) and may be used in a nationalist way with the aim of concentrating the country's efforts to its own advancement. Historically, the doctrine of isolationism can be traced back to the eighteenth century when the United States was determined to stay away from the great politics in Europe—this however ended in the 1890s when the country could no longer ignore the negative effects of the isolationism policy on its security.[127]

---

[127] For a detailed discussion on isolationism, see Klein, S., 2017. *America first? Isolationism in US foreign policy from the 19th to the 21st century* (Doctoral dissertation, Monterey, CA: Naval Postgraduate School).

Although the isolationism policy is commonly associated with the political sector, economically and specifically in the energy sector, this can apply where a country declines to cooperate with its neighbours in the enhancement of the regional energy sector by for instance refusing to pool resources for regional energy infrastructure establishment and development. It would indeed be rare for a country to apply the isolationism policy in the energy sector, and this is due to the fact that the sector not only affect the economic situation but also the politics of any given country. However, indirectly the country may refuse to cooperate by taking slow steps to respond to regional calls advocating for regional cooperation in the energy sector.

The main advantage of regionalism over isolationism in the energy sector is based on the reality that SSA countries are faced with scarce public resources and limited aid in the energy sector, and to this end these countries need to cope with their investment needs through cooperation. Moreover, some countries are more endowed with energy resources than others, so again there is need for cooperation in order to efficiently utilise the resources to benefit neighbouring countries in the region. Taking the example of the DRC, it has a large hydro potential which is estimated to be sufficient to provide three times as much power as Africa currently consumes.[128] There is therefore a need for regional cooperation to develop and utilise the available energy resources: and in this regard regional cooperation has also been recognised as being a major component of Africa's vision for its future.[129]

3.3.2 Regionalism Versus Globalism/Multilateralism

Regionalisation and globalisation have been a marked feature of the twenty-first century, changing not only the world order but also weakening the aspects of nation-state in both the political and economic spheres. It is with no doubt that regionalism in whatever form it takes inevitably imposes some constraints on state autonomy and sovereignty, and as such

[128] Auriol, E. and Biancini, S., 2015. Powering up developing countries through integration? *The World Bank Economic Review, 29*(1), pp. 1–40.

[129] International Energy Agency: World Energy Outlook Special Report, 2014. Page 179.

the nation-state is perceived as an anachronism facing forces which it can no longer control.[130] Nevertheless, it has been argued that politics and states remain vital to the efforts towards not only globalisation but also regionalism.[131]

Though this book focuses on regionalism, it is also worth noting that regionalism and globalisation are often linked together. For instance, Telò argues that both regionalism and globalisation are two components of the same historical process of strengthening interdependence and weakening the state's barriers to free trade.[132] Gamble is however quick to note that the concept of globalisation is not a new theory as global economic forces and global markets which constitute globalisation have been in existence since the emergence of capitalism.[133] Basically, the global economy as described by Gamble is understood as one in which the fundamental units are not nation states and national economies but patterns of production and consumption organised by transnational borders, and not reliant on any particular national territory or government.[134] With globalisation comes the threat to state autonomy, and it is against this background that states have embraced regionalism as a common political response to globalisation.

When compared to globalism, it has also been observed that the reality of a contemporary world seems to be better expressed in terms of regionalism than globalism.[135] Furthermore, the proponents of regional cooperation identify with its effectiveness in solving new challenges which are in most cases ignored at the global level. This argument may hold true

---

[130] Ravenhill, J., 2016. Regional integration in Africa: Theory and practice. In *Region-building in Africa* (pp. 37–52). Palgrave Macmillan US; Padoan, P.C., 2001. Political Economy of New Regionalism and World Governance. In MarioTelò, *European Union and new regionalism: Regional actors and global governance in a post-hegemonic era* (pp. 39–59). Burlington: Ashgate, at 22.

[131] Padoan, P.C., 2001. Political economy of new regionalism and world governance. In MarioTelò, *European Union and new regionalism: Regional actors and global governance in a post-hegemonic era* (pp. 39–59). Burlington: Ashgate, at 23.

[132] Telò, M. ed., 2013. *European Union and new regionalism: Regional actors and global governance in a post-hegemonic era*. Ashgate Publishing, Ltd. Page 1.

[133] Padoan, P.C., 2001. Political economy of new regionalism and world governance. In MarioTelò, *European Union and new regionalism: Regional actors and global governance in a post-hegemonic era* (pp. 39–59). Burlington: Ashgate, at 23.

[134] Ibid. Page 22.

[135] Telò, M. ed., 2013. *European Union and new regionalism: Regional actors and global governance in a post-hegemonic era*. Ashgate Publishing, Ltd.

when comparing regionalism and globalism, for instance the seriousness of certain problems such as political disintegration or epidemics may be hard to disregard at a regional level since many regional actors are immediately affected when compared to global actors.[136] In this case, considering the fact that the challenge of energy access is not a global one but rather a regional one, then regionalism is more suited to tackle such a challenge. Moreover, the lack of politically grounded problem solving at the global level also makes the regional cooperation mechanism a far better option than globalism.[137] We should however not ignore the growing interest in solving energy-related investment disputes using international or global avenues and internationally recognised investment treaties, as such a brief overview of globalism in the lens of international investment treaties is discussed in the next section, with the importance of highlighting the significance of globalism in the energy sector.

3.3.2.1 Globalism Through the Lens of International Investment Agreements

SSA is a home of energy resources, which have attracted massive investments in various parts of the region, including countries such as Somalia and South Sudan, which were initially considered too risky and politically unstable. Typically, international energy agreements are signed between the host state and the foreign investors, to ensure the effective development and exploitation of the energy resources. Professor Peter Cameron rightly summarises the interests of both the international investors and host states as follows, "*for the investor the core objective is to repatriate capital and obtain a return commensurate with, among other things, the magnitude of its investment and risk. For the host state the main goal is to obtain its agreed share of the revenue from the project and any related developments.*"[138]

---

[136] Hettne, B. and Söderbaum, F., 2006. Regional cooperation: A tool for addressing regional and global challenges. In *Meeting global challenges: International cooperation in the national interest. Final report* (pp. 179–244).

[137] Ibid.

[138] See Cameron, P., 2009. *International energy investment law: The pursuit of stability* (p. 4). Oxford University Press.

Disputes among the parties are inevitable and in this regard a significant number of African energy-related cases have been brought under or resolved through international arbitration. Some of these cases include Total E&P Uganda BV V Republic of Uganda (ICSID Case No ARB/15/11), concerning claims arising out of an allegedly unlawful tax levied by the government; Texaco Overseas Petroleum Company V The Government of the Libyan Arab Republic, AD HOC Award of January 19, 1977, concerning the nationalisation measures taken by the Libyan government; and AGIP Company v People's Republic of the Congo, ICSID Case No. ARB/77/1.

In the distant past, national legislations and judiciaries provided the regulatory investment protection structures: however, with the increasing political risks in various countries (political risks include risk resulting from unstable governance; new and adverse legislation; and adverse contract changes imposed unilaterally),[139] both contract and treaty based protection have become more acceptable options, as was illustrated in the case of Ahmadou Sadio Diallo (Republic of Guinea v Democratic Republic of the Congo).

In the case of Ahmadou *Sadio Diallo (Republic of Guinea v Democratic Republic of the Congo), the Republic of Guinea instituted proceedings against the Democratic Republic of the Congo in respect of a dispute concerning "serious violations of international law" allegedly committed upon the person of a Guinean national, Ahmadou Sadio Diallo. In response to the preliminary objection which was raised by DRC on the ground that neither of the parties had exhausted the available local remedies, the ICJ emphasised the growing role of international law and in this regard noted that*

> *The Court is bound to note that, in contemporary international law, the protection of the rights of companies and the rights of their shareholders, and the settlement of the associated disputes, are essentially governed by bilateral or multilateral agreements for the protection of foreign investments, such as treaties and the Washington Convention of 18 March 1965 on the Settlement of Investment Disputes between States and Nationals of other States, which created an International Centre for Settlement of Investment Disputes (ICSID),*

[139] See Cameron, P., 2009. *International energy investment law: The pursuit of stability* (p. 146). Oxford University Press.

> *and also by contracts between States and foreign investors. In that context, the role of diplomatic protection some-what faded, as in practice recourse is only made to it in rare cases where treaty regimes do not exist or have proved inoperative. It is in this particular and relatively limited context that the question of protection by substitution might be raised. The theory of protection by substitution seeks indeed to offer protection to the foreign shareholders of a company who could not rely on the benefit of an international treaty and to whom no other remedy is available, the allegedly unlawful acts having been committed against the company by the State of its nationality. Protection by "substitution" would therefore appear to constitute the very last resort for the protection of foreign investments.*[140]

This section will therefore focus on the treaty based investment protection, including the International Investment Agreements (IIAs) which are commonly used by developing countries to attract foreign investments.[141] Moreover, IIAs not only provide protection but also provide neutral avenues for lodging claims (mostly through international arbitration). However, there are arguments and suggestions that arbitral tribunals besides applying international law should also give effect to sustainable development principles.[142] IIAs comprise of Bilateral Investment Treaties (BITs), Multilateral Investment Treaties (MITs), and Free Trade Agreements (FTAs). IIAs basically provide for the resolving of disputes through arbitration under the rules of known international arbitration institutions including the International Chamber for Settlement of Investment Disputes (ICSID), the London Court of International Arbitration (LCIA), the International Chamber of Commerce (ICC), and the American Arbitration Association (AAA), just to mention a few.

Although many African energy rich countries are embracing international arbitration and IIAs, we cannot ignore the increasing criticism

---

[140] See para 88 of the Preliminary Objections, 24 May 2007. It can be accessed at http://www.icj-cij.org/files/case-related/103/103-20070524-JUD-01-00-EN.pdf

[141] See Cameron, P., 2009. *International energy investment law: The pursuit of stability* (p. 146). Oxford University Press.

[142] For a full discussion on the issue of international arbitration and sustainable development, see Karamanian, S.L., 2016. International energy trade and investor-state arbitration: What role for sustainable development? In *Emerging issues in sustainable development* (pp. 355–376). Springer Japan.

for both BITs and MITs. In South Africa for instance, in July 2010, a review of the BITs was conducted which enabled the South African government to make changes in the country's investment sector including cancelling BITs with several European countries.[143] Moreover, several Latin American countries including Bolivia, Ecuador, and Venezuela have in recent years announced their withdrawal from the ICSID Convention, and a regional institution with the same effect has been suggested.[144]

Basically, bilateralism as the name suggests involves two states while multilateralism involves three or more states. In arguing for multilateralism, it has been described as "*an institutional form that coordinates relations among three or more states on the basis of generalized principles of conduct: principles that specify appropriate conduct for a class of actions, without regard to the particularistic interest of the parties or the strategic exigencies that may exist in any specific occurrence*".[145] In further support for multilateralism, it has also been highlighted that bilateralism tends to favour the most powerful state and yet multilateralism favours all the parties, reasoning that "benefits from international cooperation are expected to derive over time without advance knowledge of the actors that will benefit".[146]

The need for uniformity in international investment law is also advocated so as to avoid the inconsistencies and conflicting decisions concerning similar issues.[147] A good example of an MIT in the energy sector is the Energy Charter Treaty, and this will be discussed here, specifically looking at the investment protection provisions and arbitration provisions.

[143] At the beginning of October 2012, South Africa cancelled its BITs with Belgium–Luxembourg, Spain, Germany, Switzerland, the Netherlands, and Denmark. Following the review of the BITs, the South African cabinet also decided to refrain from entering into BITs in the future, unless deemed necessary and develop a new investment act to codify typical BIT provisions into domestic law.

[144] For a full discussion on this see Vincentelli, I.A., 2010. The uncertain future of ICSID in Latin America. *Law and Business Review of the Americas, 16*, p. 409.

[145] See Schill, S.W., 2015. Multilateralization: An ordering paradigm for international investment law.

[146] Ibid.

[147] Ibid.

Energy Charter Treaty and International Energy Charter

The Energy Charter Treaty (ECT) is one of the unique multilateral instruments in the energy sector. The origins of the ECT need to be understood in the European countries' interest to trade with countries in the East following the collapse of the Soviet Union in 1991. Although the ECT was initially covering a limited scope of countries, recent developments have seen the expansion of the principles of the ECT to various regions including Asia, Africa, and Latin America. This expansion was made possible following the adoption of the International Energy Charter (IEC). Basically, the IEC is a political declaration on energy cooperation which does not bear any legal or financial obligation, and it has been adopted and signed by over 80 countries and organisations including among others the United States, China, Burundi, Uganda, Chad, ECOWAS, EAC, and the EU.

The ECT, on the other hand, is a legally binding text providing rules in the areas of investments, trade and transit, and energy efficiency. It provides dispute resolution mechanisms, while explicitly recognising and protecting national sovereignty over natural resources.[148] The purpose of the ECT as provided for in its provisions is to promote cooperation in the energy field.[149] The ECT has for more than a decade attracted a lot of studies from different scholars, and this can be attributed to its unique focus on the energy sector, one which is at the heart of every country's economic development.[150]

The recent developments in the renewable energy sector have also increased the relevance of the ECT. Notably, photovoltaic (PV) energy

---

[148] Article 18 of the Energy Charter Treaty provides for national sovereignty over energy resources. The Energy Charter Treaty and the Energy Charter Protocol on Energy Efficiency and Related Environmental Aspects were signed in December 1994 and entered into force in April 1998.

[149] See Article 2 of the Energy Charter Treaty.

[150] Some of the literature on the ECT include Konoplyanik, A. and Walde, T., 2006. Energy Charter Treaty and its role in international energy. *Journal of Energy & Natural Resources Law, 24*, p. 523; Walde, T. ed., 1996. *The Energy Charter Treaty: An east-west gateway for investment and trade* (Vol. 10). Kluwer Law International; Hobér, K., 2010. Investment arbitration and the Energy Charter Treaty. *Journal of International Dispute Settlement, 1*(1), pp. 153–190; Taheri, S. and Sharify, S.K., 2016. An analysis of the method of dispute settlement in Energy Charter Treaty compared with World Trade Organization. *Journal of Fundamental and Applied Sciences, 8*(2S), pp. 3669–3687.

investors instituted claims against some of the ECT member states (such as Italy, Spain, and Czech Republic). These claims were as a result of the domestic legislative revisions of incentive schemes for renewable energy (initially renewable energy and PV energy generation sector had benefited from large national incentive schemes and many EU countries subsidised the sector).[151] Some of the claims include that the reforms of some national supporting schemes are in breach of the fair and equitable treatment standard and ECT treaty provisions prohibiting indirect expropriation. This claim was based on the fact that the reforms had the effect of making their businesses less profitable, contrary to pre-reform legislative and regulatory commitments.

There was a general view that these renewable claims prompted Italy's withdrawal from the ECT. However, it has been highlighted that the real reasons behind Italy's withdrawal from the ECT, which she argues, are purely political as they relate to the final decision of Russia not to ratify the ECT. Besides this political reason, another reason highlighted by the Italian authorities related to the booming of intra-EU cases based on the ECT, which they considered to be clashing with the fundamental and systematic principles of EU law. The same reason was highlighted for Italy's unilateral termination of its BITs with other member states in 2009.[152]

It can be argued that Russia and Italy's withdrawal from the ECT is a major threat to both the notions of regionalism and globalism, and this has the potential of negatively affecting the ECT objective of expanding to other continents such as Africa and America. There is no proof that this will happen but we also have to account for the fact that some regional and national laws in Africa provide similar protection to investors, as that offered by ECT. For instance, the 2003 ECOWAS Protocol on Energy has similar provisions like those of the ECT.

Although some of the principles contained in the ECT such as transparency in the energy sector, open energy markets, investment protection, and transfer of technology are essential for ensuring regulatory stability, which is a prerequisite for attracting energy investors, it is worth

[151] For a full discussion on this, see De Luca, A., 2015. Renewable energy in the EU, the Energy Charter Treaty, and Italy's withdrawal therefrom.

[152] Ibid.

noting the fact that most of the founding members and signatories to the ECT are developed countries most of which have attained 100% electrification and as such are not facing energy access challenges but rather energy security challenges. This therefore implies that these global energy instruments will not have tailored energy access solutions needed in SSA, thus making regional solutions to such common regional challenges more suitable. Nevertheless, these global energy instruments also recognise the need for regional cooperation in the development of the energy sector.

3.3.3 Other Benefits of Regional Cooperation in the Energy Sector

Moving forward from ECT and IIAs, Schiff and Winters, proponents of the notion of regional cooperation, generally highlight the necessity of regional cooperation especially among neighbouring countries in the areas of public goods such as water basins, infrastructure, energy, and environment.[153]

To fully appreciate the importance of regional cooperation in the energy sector, it is essential to understand the meaning of energy cooperation. Energy cooperation has been defined to be more concerned with the provision of energy services through improved management of international energy markets, wider availability and higher quality of energy resources, and the expansion of choices in order to achieve sustainable growth: it thus incorporates the notion of a reliable ample supply of energy resources, ranging from fossil fuels to renewables and nuclear power.[154] From the above definition therefore, the potential benefits of regional cooperation are expected to be lower costs, political trust, economic development, environmental improvements, and transfer of technology among others.[155] It is also generally agreed that sustainable energy supply will be realised by cooperation.

[153] Schiff, M. and Winters, L.A. 2002. Regional cooperation, and the role of international organizations and regional integration. *World Bank Policy Research Working Paper (2872)*.

[154] Choo, J., 2006. Energy cooperation problems in Northeast Asia: Unfolding the reality. *East Asia, 23*(3), pp. 91–106.

[155] Ibid.

3.4 Barriers to Regional Cooperation

Despite the potential benefits of regional energy cooperation, there are various barriers to achieve this in SSA, including lack of infrastructure; low levels of cross-border cooperation; weak legal and institutional frameworks for energy both at national and regional levels; lack of finances; and lack of trust among member states of regional organisations.

Another difficulty of regional cooperation highlighted by some scholars is that regional cooperation agreements are typically harder to achieve than national ones because given the absence of courts or higher authorities to which to appeal, the enforcement of property rights is ambiguous and weak at the regional level.[156] However, this barrier may not be so visible in most SSA because most of them have an enforcing mechanism; for instance, the East African Community and the Economic Community of West African States have regional courts which are very functional. The Southern African Community on the other hand is in the process of establishing a functional regional Tribunal.

In response to some of the challenges experienced at a regional level, some literature highlights that the solutions to some of the difficulties in regional cooperation can be enhanced by international organisations such as the World Bank and United Nations; as these organisations can help achieve agreements that might not have been possible otherwise.[157] More still, they can use their financial resources, credibility, technical expertise, broader perspective, and neutrality to broker and enforce deals outside the set that is feasible for the countries or regions acting alone.[158] Whether the credibility of financial and international institutions can be trusted is a question worth raising, given the exploitative mechanisms employed by some of these financial institutions on the African continent. However, this is beyond the scope of this book; the focus here is on how regional organisations can be used to foster greater

---

[156] Schiff, M. and Winters, L. A., 2002. Regional cooperation, and the role of international organizations and regional integration. *World Bank Policy Research Working Paper (2872).*

[157] Ibid.

[158] Bazilian, S.T.M., 2017. The role of international institutions in fostering sub-Saharan Africa's electrification.

regional cooperation in the SSA energy sector as a way of tackling the challenge of energy access.

3.5 Dilemma Between Regional Cooperation and Energy Access

Energy access is a common challenge in SSA which requires common and cooperative solutions, and this can be achieved through the framework of the existing regional organisations. Whereas there is a clear need on the one hand for SSA countries to address the challenge of energy access, on the other hand, the same countries want to not only retain strong national sovereignty over their energy resource, but also reduce the influence of regional institutions in the energy sector.

This situation therefore makes regional cooperative mechanisms in the energy sector hard to achieve. In this section, to address the link between energy access and regional cooperation, a comparative analysis will be employed to illustrate how the two concepts of energy access and energy security have reflected the notion of regional cooperation. In doing so, regional blocks will also be discussed in a comparative manner to illustrate how they have responded to the concept of energy access.

In contrast to the concept of energy access, which only recently gained momentum in the academic sphere, a similar concept, that of energy security, has been addressed since the 1960s, mainly looking at the consumption of oil.[159] Whereas in the past, energy security was primarily associated with oil supply, now with the increasing complexity of energy systems the term has expanded. The International Energy Agency in this regard defines energy security as "the uninterrupted availability of energy sources at an affordable price".[160] The concept of energy security gained momentum in the 2000s, fuelled by an exceedingly tight and unpredictable oil market together with suggestions by some scholars of the need to expand the con-

[159] Lubell, H., 1961. Security of supply and energy policy in Western Europe. *World Politics, 13*(03), pp. 400–422.

[160] International Energy Agency, https://www.iea.org/topics/energysecurity/subtopics/whatisenergysecurity/, last visited. 20th April 2018

cept of energy security to include new factors.[161] Among the academic attention received was the idea of linking the notions of energy security and regional cooperation. For instance, some scholars connected the notions of energy security with that of regional cooperation by putting forward arguments to the effect that energy security does not stand by itself but rather is lodged in the larger relationships among nations and how they interact with one another.[162] This argument is amplified in recent literature, which appears to suggest that states should explicitly include and refer to regional cooperation mechanisms in their national energy policies and strategies so as to enhance regional energy security.[163]

Comparatively, there has been little academic reflection on the link between regional cooperation and the concept of energy access. Some scholars have nonetheless given some thought to the link by including new subjects such as climate change. Uddin and Taplin, for instance, seem to suggest that regional cooperation initiatives linked with climate change mitigation objectives could assist in widening energy access.[164]

A comparative analysis is also employed in this chapter to highlight the practical steps taken under the auspices of regional cooperation to tackle challenges of energy security and energy access. When it comes to the issue of energy security, there are practical regional cooperation mechanisms established in the form of inter-governmental organisations to respond to the challenge of energy security. The Organization of the Petroleum Exporting Countries (OPEC)[165] and the IEA[166] are good examples of the link between energy security and regional cooperation.

[161] Cherp, A. and Jewell, J., 2014. The concept of energy security: Beyond the four as. *Energy Policy, 75*, pp. 415–421.

[162] Yergin, D., 2006. Ensuring energy security. *Foreign Affairs*, pp. 69–82.

[163] Choo, J., 2006. Energy cooperation problems in Northeast Asia: Unfolding the reality. *East Asia, 23*(3), pp. 91–106.

[164] Uddin, N. and Taplin, R., 2015. Regional cooperation in widening energy access and also mitigating climate change: Current programs and future potential. *Global Environmental Change, 35*, pp. 497–504.

[165] OPEC is a permanent inter-governmental organisation created at the Baghdad Conference on 10–14 September 1960. It comprises of 14 member states including: Iran, Iraq, Kuwait, Saudi Arabia, Venezuela, Indonesia (suspended its membership in November 2016), Libya, United Arab Emirates, Algeria, Nigeria, Ecuador, Angola, and Gabon.

[166] IEA member states include Australia, Austria, Belgium, Canada, Czech Republic, Denmark, Estonia, Finland, France, Germany, Greece, Hungary, Ireland, Italy, Japan, South Korea,

Looking briefly at OPEC, one can notice the historical shift with regard to cooperation in the energy sector. Historically, the main objective underlying the establishment of OPEC in the 1960s, as stipulated in the Declaratory Statement, was to focus on the inalienable right of all countries to exercise permanent sovereignty over their natural resources in the interest of their national development. The Declaratory Statement of Petroleum in Member Countries was adopted in 1968 and it rhymes with the situation at the time, as during the 1960s there was extensive decolonisation and the birth of many new independent states in the developing world. The mandate of the Statement was however expanded in the next decades taking into account energy security issues. To date, OPEC plays an influential role in unifying petroleum policies among member states, in addition to securing fair and stable prices for petroleum producers.[167]

The IEA, on the other hand, was initially established in the wake of the 1973–1974 oil crisis to respond to physical disruptions in the supply of oil as well as serving as an information source on statistics about the international oil market and other sectors. The establishment of the IEA illustrated a strong link between regional cooperation and energy security. It was established under the framework of the OECD, notwithstanding the existence of formal OECD energy structures which could deal with pressing energy matters.

The establishment of OPEC and the IEA therefore reflects the relationship between regional cooperation and energy security. The continued progress of these inter-governmental organisations illustrates and supports the argument in this book that it is essential to understand and get the problem right if we are to find binding and viable solutions to such problems. In this regard, it is essential to fully understand the concept of energy access to avoid misconceived responses to the challenge,

Luxembourg, the Netherlands, New Zealand, Norway, Poland, Portugal, Slovakia, Spain, Sweden, Switzerland, Turkey, the United Kingdom, and the United States. Only OECD member states are members on the IEA.

[167] For more information on the historical transformation of OPEC, look at the Organisation of the Petroleum Exporting Countries (OPEC) website at http://www.opec.org/opec_web/en/about_us/24.htm, last visited on 3 January 2017.

and to provide specific solutions. Comparatively, when it comes to the concept of energy access, we are yet to witness an inter-governmental organisation formed specifically to tackle this challenge.

Moving forward and examining some of the forums of regional cooperation in SSA such as regional organisations, we notice the lack of specifically governing legal energy instruments to address energy access. Taking the EAC as an example, in spite of its pivotal role in spearheading regional cooperation mechanisms in the energy sector, the EAC has neither a regional energy protocol nor a regional energy policy. Moreover, the organisation has only two energy officials within the energy department. The EAC only refers to energy generally under Article 101 of the Treaty for the establishment of EAC, stipulating that, "*The Partner States shall adopt policies and mechanisms to promote the efficient exploitation, development, joint research and utilization of various energy resources available within the region.*"[168] Despite the fact that the Treaty was enacted in 1999, as of yet no energy policy, energy protocol, or even energy Act has been adopted at a regional level. This therefore shows the level of reluctance at the EAC regional level to legislate on energy matters, and yet other sectors like gender have been given more attention.[169]

The recommendation in this regard is for EAC to use its legislative capacity to enact an Energy Access Bill (later to become a legally binding Energy Act). This Bill should emphasise the need for regional energy cooperation including making a provision for the autonomy and independence of regional energy institutions in the EAC. Once enacted into law, the EAC Energy Act should be adopted by the member states and it should also take into consideration new technological developments in renewable energy. The energy legislation in EAC can take either the form of an Act or an energy protocol, with the main objective of setting common principles, rules, and regulations governing the energy sector in East

[168] Article 101 (1) of The Treaty for the Establishment of the East African Community, 1999.

[169] Since 1999, EAC has taken advantage of its legislative organ the East African Legislative Assembly (EALA) to enact laws to overcome challenges in other sectors. These include among others the EAC Appropriation Act of 2012, The EAC Vehicle Load Control Act of 2013, and recently the EAC Community One Stop Border Posts Act of 2016. There are also Bills such as the EAC Prohibition of Female Genital Mutilation Bill of 2016, the EAC Human and Peoples Rights Bill of 2011, and the EAC Holiday Bill of 2013, just to mention but a few.

Africa. Comparatively, other SSA regional organisations such as ECOWAS and SADC have specific energy protocols.

3.6 Conclusion

In conclusion, the notion of regional cooperation has a lot to offer in the twenty-first century. However, its survival is questionable. Moreover, for this notion to be successful in tackling the challenge of energy access, there is a need to first precisely understand what the problem is and come up with focused regional initiatives to tackle this problem. There is also a need to find specific regional mechanisms dedicated to energy access. In Chaps. 4, 5, and 6, we explore some examples of regional cooperation mechanisms that have been employed in SSA to tackle this challenge and analyse their effectiveness.

References

Allies in adversity: The frontline states in Southern African security, 1975–1993. Ohio University Press.

Auriol, E., & Biancini, S. (2015). Powering up developing countries through integration? *The World Bank Economic Review, 29*(1), 1–40.

Bazilian, M., & Tagliapietra, S. (2017). The role of international institutions in fostering sub-Saharan Africa's electrification.

Botswana Power Corporation Act, 1970 (Chapter 74:01).

Cameron, P. (2009). *International energy investment law: The pursuit of stability* (p. 146). Oxford University Press.

Cherp, A., & Jewell, J. (2014). The concept of energy security: Beyond the four as. *Energy Policy, 75*, 415–421.

Choo, J. (2006). Energy cooperation problems in Northeast Asia: Unfolding the reality. *East Asia, 23*(3), 91–106.

De Luca, A. (2015). Renewable energy in the EU, the Energy Charter Treaty, and Italy's withdrawal therefrom. *Oil, Gas & Energy Law Journal (OGEL), 13*(3).

Electricity Supply Amendment Act, Chapter 73:01.

Ethier, W. (1998). *The international commercial system (No. 210)* (p. 11). Princeton University International Economics.

Hettne, B., & Söderbaum, F. (2006). Regional cooperation: A tool for addressing regional and global challenges. In *Meeting global challenges: International cooperation in the national interest. Final report* (pp. 179–244).

International Energy Agency: World Energy Outlook Special Report, 2014.

Karamanian, S. L. (2016). International energy trade and investor-state arbitration: What role for sustainable development? In *Emerging issues in sustainable development* (pp. 355–376). Springer Japan.

Klein, S. (2017). *America first? Isolationism in US foreign policy from the 19th to the 21st century* (Doctoral dissertation). Monterey, CA: Naval Postgraduate School.

Levine, D. H., & Nagar, D. (Eds.). (2016). *Region-building in Africa: Political and economic challenges*. Springer.

Lubell, H. (1961). Security of supply and energy policy in Western Europe. *World Politics, 13*(3), 400–422.

MarioTelò, European Union and new regionalism: *Regional actors and global governance in a post-hegemonic era* (pp. 39–59) Burlington: Ashgate, at 25.

Nkrumah, K., Arrigoni, R., & Napolitano, G. (1963). *Africa must unite* (p. 97). London: Heinemann. p. 133.

Nye, J. S. (Ed.). (1968). *International regionalism: Readings*. Little, Brown.

Nyerere, J. K. (1966). The dilemma of the Pan-Africanist. *Address to the University of Zambia Congregation, 13* (pp. 207–217).

Omari, A. H., & Macaringue, P. (2007). Southern African security in historical perspective. *Security and democracy in Southern Africa*; Khadiagala, G.M., 1994.

Padoan, P. C. (2001a). Political economy of new regionalism and world governance. In M. Telò (Ed.), *European Union and new regionalism: Regional actors and global governance in a post-hegemonic era* (pp. 39–59). Burlington: Ashgate, at 22.

Padoan, P. C. (2001b). Political economy of new regionalism and world governance. In M. Telò (Ed.), *European Union and new regionalism: Regional actors and global governance in a post-hegemonic era* (pp. 39–59). Burlington: Ashgate, at 23.

Ravenhill, J. (2016). Regional integration in Africa: Theory and practice. In *Region-building in Africa* (pp. 37–52). Palgrave Macmillan US.

SADC Protocol on Energy, 1996.

SADC Renewable Energy and Energy Efficiency Status Report, 2015.

Schiff, M., & Winters, L. A. (2002). Regional cooperation, and the role of international organizations and regional integration. *World Bank Policy Research Working Paper (2872).*

Schill, S. W. (2015). Multilateralization: An ordering paradigm for international investment law.

Telò, M. (Ed.). (2013). *European Union and new regionalism: Regional actors and global governance in a post-hegemonic era*. Ashgate Publishing, Ltd.

The 1992 SADC Treaty.

The ECOWAS Energy Protocol, 2003.

The Swaziland Investment Promotion Act, 1998.

The Zimbabwe Investment Authority Act (Chapter 14:30).

Treaty for the Establishment of the East African Community, 1999.

Uddin, N., & Taplin, R. (2015). Regional cooperation in widening energy access and also mitigating climate change: Current programs and future potential. *Global Environmental Change, 35*, 497–504.

Vincentelli, I. A. (2010). The uncertain future of ICSID in Latin America. *Law and Business Review of the Americas, 16*, 409.

Yergin, D. (2006). Ensuring energy security. *Foreign Affairs*, 69–82.

4

Regional Cooperation in Renewable Energy and Fossil Fuel Development

4.1 Introduction

Sub-Saharan Africa (SSA) has vast energy resources, including both conventional and unconventional of which most are untapped. The region, for instance, has natural gas potential of approximately 503.3 Tcf.[1] However, low electrification rates coupled with heavy reliance on inefficient energy sources such as traditional biomass are rampant. The fluctuating fossil fuel prices coupled with their negative impact on the environment has led to massive investments and an increase in the development of alternative clean energy sources. Renewable Energy Sources (RES) are now widely recognised as being pivotal not only to solving SSA's energy access challenges, but also those concerning climate change.

Generally speaking, the heavy reliance on inefficient energy sources such as traditional biomass and fossil fuels negatively affects the environment and has been linked to global warming. In this regard, global efforts

[1] See BP Statistical Review of World Energy, 2017. It can be accessed at https://www.bp.com/content/dam/bp/en/corporate/pdf/energy-economics/statistical-review-2017/bp-statistical-review-of-world-energy-2017-full-report.pdf

V. R. Nalule, *Energy Poverty and Access Challenges in Sub-Saharan Africa*, Energy, Climate and the Environment, https://doi.org/10.1007/978-3-319-95402-8_4

to address global warming and climate change have been echoed by the 2015 Paris Agreement. Taking into consideration that SSA has massive energy resources such as coal and fossil fuels, and given the region's energy access challenges, the fundamental question that arises is whether these resources could be utilised in a sustainable manner to address the challenge of energy access. Further, can regional organisations enhance the cooperative development of renewable energy in addressing both energy access and climate change in SSA?

In addressing the questions raised above, the trilemma between energy access, climate change, and renewable energy will be discussed from a regional cooperation perspective, specifically looking at Regional Economic Communities (RECs). The book, through a comparative analysis, explores how regional cooperation enhances the development of renewable energy in addressing both energy access and climate change challenges in SSA. This chapter focuses on both the legal and institutional frameworks in place to support the regional development of RES in SSA. Additionally, it analyses whether international instruments such as the Paris Agreement support regional cooperation in tackling both climate change and energy access in SSA.

Taking stock of the above, a tri-step framework in terms of sections is employed to address the issues raised. Although this chapter focuses on renewable energy, the first section looks briefly at the potential of fossil fuels in addressing the challenge of energy access in SSA. This section is important given the current high reliance on fossil fuels to meet the energy demands of various countries in Africa and other regions including America, Europe, and Asia. Additionally, there has been continued discovery of oil and gas in many African countries and there are ongoing efforts to ensure that such resources contribute to addressing the challenge of energy access.

Section two discusses the trilemma and nexus between energy access, climate change, and renewable energy. This section begins by examining climate change and its impact in SSA and how this relates to the energy sector. Subsequently, the section discusses how renewables can address both climate change and energy access challenges. This section is important as it includes a discussion of the role of regional cooperation in addressing the aforementioned challenges in SSA.

Section three looks at a comparative analysis of the existing regional and international initiatives in the development of RES. Specifically, institutional and policy frameworks in EAC, SADC, and ECOWAS are discussed. The discussion of this section pinpoints the legal framework behind the establishment and development of regional renewable energy projects, which are the focus of the subsequent chapters. The concluding remarks in this chapter reflect on the issues raised.

4.2 Potential of Fossil Fuels in Addressing the Challenge of Energy Access

As noted in Chap. 1, SSA is a home of massive fossil fuels and coal resources. Just like RES, fossil fuels can play a significant role not only in increasing the electricity access rate in SSA, but also escalating industrialisation and urbanisation. In this regard, this section examines the potential of fossil fuels in addressing energy access in SSA. The first part of this section analyses the role of fossil fuels in electricity generation in SSA, pinpointing how energy access can be tackled by fossils. Part two of this section analyses the current energy demand as reported by British Petroleum (BP). This is important as it highlights the crucial role of fossils in different sectors including industry and transport—which sectors are key in ensuring the development of various countries in SSA. While discussing the role of fossil fuels, we cannot ignore their impact on the environment and communities. In this regard, the last part of this section examines the issues relating to gas flaring and oil spills in the regions where these resources are extracted from, specifically looking at the Niger Delta in Nigeria. This part is important as it opens a discussion on climate change, its relation to the energy sector, and its impact on SSA.

4.2.1 Role of Fossil Fuels in Electricity Generation in SSA

Electricity can be generated from numerous energy sources including thermal (thermal power is generated using coal and oil), hydro (hydropower is produced from water sources including construction dams over overflowing rivers), nuclear (nuclear power plants use uranium as fuel for generating power), and from renewables including solar energy, wind energy, tidal energy, and bio energy. Despite the vast energy resources, electricity deficiency is rampant in African countries. Looking at the Southern African region for instance, it has been observed that besides DRC and Mozambique, most countries in the SADC region have a supply deficit.

According to the SADC Energy Monitor, as of November 2015, the Southern Africa Power Pool (SAPP) installed generation capacity stood at 61,859 MW, although the available generation was only 46,910 MW.[2] In this section, the different sources of fossil fuels will be discussed in relation to the energy mix and electricity generation capacity in the SADC region as illustrated later.

4.2.2 SAPP Installed Generation Capacity[3]

The SAPP heavily relies on coal, which accounts for over 62% of the total generation capacity, followed by renewables including hydro at 21.02%, wind at 4.03%, solar PV at 2.94%, and distillate at 4.38%. Although traditional biomass in the form of firewood is relied on by most people in rural areas, we note that in terms of electricity generation, this has minimal capacity and as such biomass generally stands at 0.07% of the SAPP installed generation capacity.[4] Given that this section focuses on fossil fuels, the various sources are discussed below.

4.2.2.1 Coal

Coal is a major source of energy not only in Africa but also other regions including Asia, Europe, and America. It provides approximately 41% of the world's electricity needs; and global coal supply is predicted to increase

[2] See SADC Energy Monitor, 2016. Page 33.

[3] See SADC Energy Monitor, 2016. Page 33.

[4] Ibid.

at an average rate of 0.6% through 2020.[5] However, there are concerns that developed and developing countries should reduce their coal dependence for energy production and instead look to other cleaner technologies such as renewables.

In SSA, taking the example of Southern Africa, coal is the most dominant source of electricity in the SADC region, contributing to over 60% followed by hydro, which contributes 21.02% of the electricity generation capacity. This heavy reliance on coal in SAPP can be attributed to the fact that South Africa dominates the power generation as it accounts for 76% of the overall generation capacity. Moreover, as of March 2015, at least 86% of South Africa's total generation capacity of 44,170 MW came from coal fired plants, while 82% of Botswana's electricity was produced from coal, and 63% for Zimbabwe. High reliance on coal for electricity in South Africa is the main reason for the high GHG emission. It has been argued that hydroelectricity could play a big role in reducing South Africa's GHG emissions, especially given the large hydro schemes in the Congo and Mozambique, which could provide an alternative electricity source for South Africa.[6] Besides the option of hydroelectricity to replace coal, other cleaner electricity generation options such as imported natural gas feeding into combined-cycle gas turbines (CCGTs) and the pebble bed modular reactor (PBMR) have also been suggested for South Africa especially due to their low GHG emissions.[7]

Although coal is a main source for South Africa, Botswana, and Zimbabwe, on the other hand, however, other SADC countries such as DRC, Lesotho, Malawi, and Zambia solely rely on hydropower as a source of electricity generation. Recent criticism over the use of coal has encouraged investments in other energy sources including oil, gas, and renewables, in the generation of electricity. There are also plans for new clean coal technologies such as supercritical, fluidised bed combustion, and integrated gasification combined-cycle plants.

---

[5] See International Energy Agency, https://www.iea.org/about/faqs/coal/, last visited on 4 March 2018.

[6] See Mukheibir, P., 2017. Possible climate change impacts on large hydroelectricity schemes in Southern Africa. *Journal of Energy in Southern Africa, 18*(1), pp. 4–9.

[7] Ibid.

4.2.2.2 Oil Resources

As elaborated in the previous chapters, Africa is a home of massive oil resources. Generally, oil is considered as the most important source of energy as it is used in automobiles, planes, trains, and ships among many other uses. In terms of access to energy, electricity is also generated from distillate power plants which basically generate electricity using diesel fuel in countries such as Mozambique, Namibia, and Tanzania and it accounts for close to 5% of the total generation. In this regard oil resources can contribute in addressing the challenge of energy poverty.[8]

Despite massive oil and gas resources, Africa remains a net importer of petroleum products, necessitating the need to invest in more oil refineries. Comparatively, the United States with an history of petroleum refinery dating back as far as 1850, as of January 2015, had 137 operating refineries. These US refineries far exceed in number the 46 refineries found on the entire African continent. The investments in petroleum refineries which have been embraced by the United States and other developed countries have indeed contributed to their export capacity in refined products. For instance, in 2013, the United States produced 18.9 million barrels per day of refined petroleum products, more than any other country. There is therefore a need for African countries to invest in more refineries, as this will ensure that these countries are less exposed to market fluctuation and manipulation and also these countries will be able to retain control of supply and marketing of petroleum products. Most importantly, petroleum resources if utilised can help in solving energy access challenges in SSA.

4.2.2.3 Natural Gas

Natural gas is another form of energy that can play a role in addressing energy access challenges in SSA. Moreover, it is also used in fertiliser and petrochemical plants and gas based thermal power plants.

[8] See SADC Energy Monitor, 2016. Page 33.

Natural gas can be produced in two ways: with petroleum products as associated gas; or free gas, which is obtained from gas fields. In terms of electricity generation, natural gas can be used in the combined-cycle gas: Combined-Cycle Gas Turbines (CCGT) are a form of highly efficient energy generation technology that combines a gas-fired turbine with a steam turbine. The design uses a gas turbine to create electricity and then captures the resulting waste heat to create steam, which in turn drives a steam turbine, significantly increasing the system's power output without any increase in fuel. The technology is typically powered using natural gas, but it can also be fuelled using coal, biomass, and even solar power as part of solar combined-cycle plants.[9]

Nuclear and combined-cycle gas turbine power generators are not common in SADC and as such these contribute to 3.3% and 1.3% of the total electricity generation, respectively. In SADC it is only South Africa which has a nuclear power station—Koeberg, which contains two uranium pressurised water reactors.[10] Combined-cycle gas turbine power plants are also not common and can only be traced in DRC, Malawi, and Tanzania.[11]

4.2.3 Influence of Fossil Fuels on Other Sectors

In the previous section, we explored the influence of fossil fuels on electricity generation, a major area of concern in SSA. Besides electricity generation, fossil fuels are also in high demand in other sectors including industry and transport. This section therefore examines the demand and consumption of various energy sources as illustrated in the BP report, specifically highlighting the increasing demand for fossil fuels in various sectors of the economy.

---

[9] See BusinessGreen, https://www.businessgreen.com/bg/glossary/1807455/combined-cycle-gas-turbine-ccgt, last visited on 4 March 2018.

[10] See SADC Energy Monitor, 2016. Page 35.

[11] Ibid.

4.2.3.1 Analysis of the 2018 BP Energy Outlook

According to the 2018 BP Energy Outlook, there is currently a high demand for both fossil fuels and renewables in different sectors and this is expected to grow by around a third by 2040.[12] The increase in the global energy demand is mainly driven by increasing prosperity in fast-growing emerging economies such as China and India. This increase is also supported by population growth, estimated to increase by around 1.7 billion to reach nearly 9.2 billion people in 2040. Additionally, the global boom in urbanisation is projected to increase, as almost 2 billion people are likely to live in urban centres by 2040 and Africa is projected to contribute one-third of this increase in urbanisation: Productivity levels are also expected to increase and it is estimated that 2.5 billion people will be lifted from low incomes. However, the African continent is anticipated to be weighed down by weak productivity, and as such it is anticipated to account for over half of the increase in global population but less than 10% of world GDP growth.[13]

Reflecting on the discussion above, we note that Africa will require more energy to meet the anticipated urbanisation and as such fossil fuels will be on a high demand. This therefore can have the potential of escalating the challenge of energy access, especially considering the fact that the increasing energy demand might not match with the supply. The graph below illustrates how this energy transition specifically in the demand side will look as predicated by BP (see Fig. 4.1).

As illustrated in the graph above, unlike fossil fuels which have been on a high demand for decades, the demand for renewables on the other hand was very minimal in the years preceding 2010. The BP report however projects an increase in renewable energy demand between 2020 and 2040 and as such this is anticipated to account for over 40% of the global energy demand by 2040. Notwithstanding the increase in renewable energy demand, fossil fuel demand is also accelerating, necessitating more investments in clean technology to utilise these resources (the demand for coal is projected to decrease following China's anticipated reduction in coal use).[14]

---

[12] See, BP Energy Outlook, 2018, https://www.bp.com/content/dam/bp/en/corporate/pdf/energy-economics/energy-outlook/bp-energy-outlook-2018.pdf ENERGY OUTLOOK, last accessed on 3 March 2018.

[13] Ibid.

[14] Ibid.

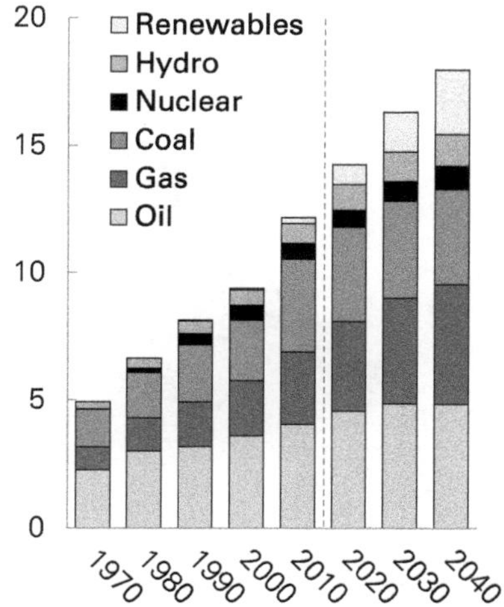

Fig. 4.1 Primary energy demand by fuel type. Source: BP Energy Outlook, 2018 (see footnote 12)

Additionally, it is essential to consider the key sectors influencing the global demand and consumption of energy. Currently, the industrial sector (including the non-combusted use of fuels) consumes around half of all the global energy and feedstock fuels, residential and commercial buildings account for 29%, transport 20%, and other sectors account for the remainder.[15] The graph below illustrates the energy demand as per these sectors (see Fig. 4.2).

In the BP Evolving Transition (ET),[16] the industrial sector is expected to account for around half of the increase in energy consumption[17]; energy use in buildings is also anticipated to increase rapidly driven by

[15] See BP Energy Outlook, 2018, https://www.bp.com/content/dam/bp/en/corporate/pdf/energy-economics/energy-outlook/bp-energy-outlook-2018.pdf ENERGY OUTLOOK, last accessed on 3 March 2018.

[16] In the BP Energy Outlook, the Evolving Transition (ET) scenario assumes that government policies, technology, and social preferences continue to evolve in a manner and speed seen over the recent past. It is however noted that carbon emission in this scenario is not consistent with achieving the Paris Climate goals. See BP Energy Outlook, 2018. Page 13. It can be accessed at https://www.bp.com/content/dam/bp/en/corporate/pdf/energy-economics/energy-outlook/bp-energy-outlook-2018.pdf ENERGY OUTLOOK, last accessed on 3 March 2018.

[17] The non-combusted use of fuels as feed stocks for petrochemicals, lubricants, and bitumen is anticipated to become an important component of the overall industrial demand over the outlook.

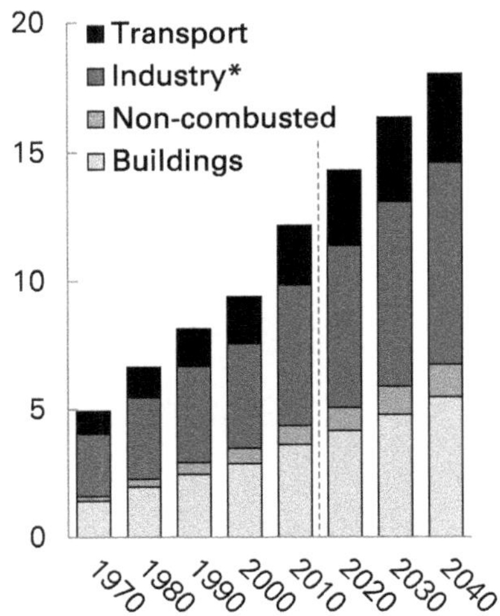

Fig. 4.2 Primary energy demand: End-use sector (billion tons). Source: BP Energy Outlook, 2018 (see footnote 15)

demand for space cooling, lighting, and electrical appliances. The transport sector however is expected to experience a decrease in energy demand, mostly attributed to the improvements in vehicle efficiency. In addition, it is estimated that the world will be more electrified and as such almost 70% of the increase in primary energy will be going to the power sector.

The discussion above highlights the importance of fossil fuels in the key developmental sectors including industry, transport, and buildings. Connecting the discussion to this book, we note that since most countries in SSA are still developing, they not only need energy for cooking and lighting (the key elements of energy access), but these countries will also require massive energy for industrialisation and urbanisation. In this regard therefore, it becomes important to invest more in clean technology to ensure sustainable utilisation of fossil fuels in Africa as a way of addressing energy access and ensuring economic development.

4.2.4 Negative Effects of Fossil Fuels

Despite the potential of fossil fuels to tackle the challenge of energy access in SSA, there are various limitations to the development of this sector including lack of exploration to increase the size of proven reserves; lack of human skills and resources; and lack of essential infrastructure such as pipelines, storage, and refining facilities.

Moreover, fossil fuels are a major contributor to climate change as they account for around 60% of total greenhouse gas emissions. In this section therefore, the negative impacts of fossil fuels will be examined, specifically looking at the issue of gas flaring and oil spills in the Niger Delta region.

Broadly, the issue of gas flaring has attracted a lot of attention and there are visible efforts to address it. However, the problem persists in various SSA countries including Angola, South Sudan, and Nigeria. Angola for instance in an effort to address gas flaring, has expanded the production of liquefied natural gas (LNG), which is used in cooking and general heating. Gas-fired power stations have also been established in countries such as Mozambique and South Africa.[18]

Nevertheless, due to lack of infrastructure and required finances to utilise gas, many countries resort to gas flaring, a method which has faced a lot of criticism from both environmental and human rights activists. This is due to the fact that it contributes significantly to greenhouse gas (GHG) emission and from an economic point of view, it is considered wasteful. Below an overview of gas flaring is provided, particularly considering Nigeria as a case study.

Gas Flaring in Nigeria

Basically, gas flaring is the burning or venting of the associated gas released during crude oil production. The issue of gas flaring in Nigeria is an important one considering the fact that the country is endowed with large reserves of both associated and non-associated natural gas. As of 2016,

[18] SADC Energy Monitor, 2016. Page 77.

Nigeria had 186.6 Tcf proven natural gas reservoirs, most of which is associated natural gas (ANG).[19]

The Niger Delta region is presently the singular deposit of oil and gas in Nigeria. Basically, when ANG is produced, it can be vented, flared, or re-injected back into the earth or sometimes captured, processed, and sold. Despite these various options, massive ANG continue to be flared in the Niger Delta and this is estimated at about 1.4 billion cubic feet. Gas flared in Nigeria stood at 20.6% in 2013, 15.28% in 2014, and 13.25% in 2015.[20]

It has been noted that gas flaring in Nigeria highlights the environmental, socio-economic, and political problems posed by resource extraction to the people of the Niger Delta region. Gas flaring has not only impoverished the communities in the Niger Delta, but also increased mortality rate and exposed the local people to strange diseases including cancer, respiratory challenge, and skin irritation. Moreover, putting Russia aside, Nigeria flares more ANG than any other country in the world.[21]

Connecting the Niger Delta situation to this chapter is the fact that the flared ANG can meet the energy needs of Nigeria and other countries in West Africa, and this definitely means that the issue of energy poverty and energy access can be adequately tackled if the flared gas is utilised in a sustainable manner. Additionally, besides tackling the issue of energy access in West Africa, it has been highlighted that converting ANG to productive use will generate employment opportunities, widen the economic base of the country, and reduce armed insurrection in the region.[22] Other scholars have further suggested a strategic fuel substitution initiative as a way of reducing gas flaring: this is to the effect that natural gas be used in the transport sector as a substitution fuel for gasoline and diesel (the transport sector is considered to be the largest consumer of oil in terms of gasoline and diesel).

The effects of gas flaring include the following: the gas flared contains toxic compounds such as methane, which have negative environmental impacts; the blazing fire emits a lethal chemical known as benzene, which causes birth damage; it also has adverse negative effects on plants. Considering the negative impacts of gas flaring, several policies aimed at

[19] See BP Statistical Review of World Energy, 2017. Page 26. It can be accessed at https://www.bp.com/content/dam/bp/en/corporate/pdf/energy-economics/statistical-review-2017/bp-statistical-review-of-world-energy-2017-full-report.pdf, last visited on 20 March 2018.

[20] Giwa, S.O., Nwaokocha, C.N. and Odufuwa, B.O., 2017. Mitigating gas flare and emission footprints via the implementation of natural gas vehicles in Nigeria. *Energy Policy*, *111*, pp. 193–203.

[21] Hassan, A. and Kouhy, R., 2013, June. Gas flaring in Nigeria: Analysis of changes in its consequent carbon emission and reporting. In *Accounting Forum* (Vol. 37, No. 2, pp. 124–134). Elsevier.

[22] Akinola, A.O., 2017. Resource mis governance and the contradictions of gas flaring in Nigeria: A theoretical conversation. *Journal of Asian and African Studies*, p. 0021909617722374.

reducing and utilising the gas have been set up in Nigeria, and these range from voluntary to compulsory efforts. For instance, as far back as 1969, the Nigerian government made efforts to reduce gas flaring and in this regard, it instructed oil and gas companies to set up infrastructure that could facilitate effective utilisation of ANG. This was not a legally backed initiative and as such it was not taken seriously by the oil and gas companies. It is on this basis that various laws were enacted including the Associated Gas Reinjection Act of 1979 and the Associated Gas Framework Agreement (negotiated between the Nigerian government and the oil and gas companies). Besides the national laws, there are other projects aimed at reducing gas flaring including the Nigerian Liquefied National Gas (NLNG) project and the West Niger Delta Liquefied Natural Gas project. In addition, the West African Gas Pipeline is also another positive development. Moreover, Nigeria also joined the Global Gas Flaring Reduction (GGFR) public private partnership.

Despite the various laws aimed at ending gas flaring, the problem persists and it has been noted that the failure of these laws to be effective is directly linked with the endemic of corruption in the Nigerian government.[23] Besides corruption, there are other barriers to achieve zero flaring and these include limited and insufficient market for gas and its related products; lack of gas infrastructure; insufficient funding and resources; low and non-competitive gas and gas product prices; and high level of costs associated with gas utilisation and reinjection schemes.[24] Although the effects of gas flaring have mainly affected the local communities in the Niger Delta regions, it important to note that pollution knows no boundaries and as such gas flaring ceases to be only a national issue, but rather it is also a regional and global challenge. This therefore necessitates more regional efforts to tackle gas flaring and ensure that the gas is utilised regionally to tackle energy poverty challenges in SSA.

Oil Spills in the Niger Delta Region

Besides gas flaring challenges in the Niger Delta region, there is also the challenge of oil spills. Basically "oil spill" is defined as an accidental release of a liquid petroleum hydrocarbon into the environment due to human activity. In Nigeria, there are several causes of oil spills, and it has been noted that approximately 50% of the spills are attributable to pipeline vandalism and tanker accidents; 28% spills are due to sabotage; production

[23] Akinola, A.O., 2017. Resource mis governance and the contradictions of gas flaring in Nigeria: A theoretical conversation. *Journal of Asian and African Studies*, p. 0021909617722374.

[24] For a full discussion see Hassan, A. and Kouhy, R., 2013, June. Gas flaring in Nigeria: Analysis of changes in its consequent carbon emission and reporting. In *Accounting Forum* (Vol. 37, No. 2, pp. 124–134). Elsevier.

operations account for 21%; and the remaining 1% is due to deficient equipment used in production. Additionally, deliberate blowing up of pipelines by militant groups has also increased oil spillage in the Niger Delta region.[25]

Generally, oil spills are categorised into three, depending on the quantity of oil spilled. There are minor spills (when quantity spilled is less than 25 barrels in inland waters such as lakes or less than 250 barrels on land or coaster water); medium spills (when the spilled volume is not more than 250 barrels in inland waters or in a range of 250–2500 barrels in coastal waters); major spills (occurs when the spilled volume is above 250 barrels in inland waters and 2500 barrels in coastal waters); catastrophic spills (volume spilled exceeds that in major spills, and as a result there is a serious threat to human welfare and health).[26]

Oil spills are very prevalent in Nigeria and it is estimated that about one million to thirteen million tons of hydrocarbons have been spilled in the region over the last 50 years. These have led to the destruction of rainforest habitat occupying a land portion of 7400 square kilometres, loss of mangrove forests, destruction of an equivalent of a year's food, and strange diseases to the local people, just to mention but a few.[27]

The issue of oil spills in the Niger Delta region of Nigeria gained a lot of international attention following the release of the Environmental Assessment of Ogoniland by the United Nations Environment Programme (UNEP) in 2011. The report among others found that there was widespread oil contamination in Ogoniland and the people have been living with this pollution for more than 40 years and continue to live with it. Additionally, the report found that the various institutions and ministries responsible for tackling the oil spills lacked the necessary resources and as such this made the enforcement of environmental protection laws very hard. Nevertheless, the report identified some emergency measures which had to be tackled immediately, considering their serious negative impact to the people in the region: These emergency measures are eight in number and they include the following: ensuring that all drinking water wells where hydrocarbons were detected are marked and that people are informed of the danger;

[25] Allison, C., Oriabure, G., Ndimele, P.E. and Shittu, J.A., 2018. Dealing with oil spill scenarios in the Niger Delta: Lessons from the past. In *The political ecology of oil and gas activities in the Nigerian aquatic ecosystem* (pp. 351–368).

[26] Ibid.

[27] Zabbey, N., Sam, K. and Onyebuchi, A.T., 2017. Remediation of contaminated lands in the Niger Delta, Nigeria. See, Allison, C., Oriabure, G., Ndimele, P.E. and Shittu, J.A., 2018. Dealing with oil spill scenarios in the Niger Delta: Lessons from the past. In *The political ecology of oil and gas activities in the Nigerian aquatic ecosystem* (pp. 351–368).

providing adequate sources of drinking water to those households whose drinking water supply is impacted; people in Nsisioken Ogale who have been consuming water with benzene over 900 times the WHO guideline are recorded on a medical registry and their health status assessed and followed up; initiate a survey of all drinking water wells around those wells where hydrocarbons were observed; post signs in areas where hydrocarbons were observed on surface water warning people not to fish, swim, or bathe in it; post signs around all the sites identified as having contamination exceeding intervention values; inform all families whose rainwater samples tested positive for hydrocarbons and advise them not to consume the water; and mount a public awareness campaign to warn the individuals who are undertaking artisanal refining that such activities are damaging their health.[28]

Following the publication of the UNEP report, some developments have occurred in this respect including the introduction by the Federal Government of Nigeria of a clean-up and restoration of Ogoniland project on the shores of Bodo Greek on 2 June 2016. Besides this project, there are several other projects that were introduced in the past even before the release of the 2011 UNEP Ogoniland Report, not to mention the various national and international laws in place. For instance, the Clean Nigeria Associates (CNA) was formed in 1981, with the aim of building strategies to curb oil pollution; environmental justice movements such as the Movement for the Survival of the Ogoni People (MOSOP) were also established under the leadership of Ken-Saro Biwa, and in 1990, MOSOP presented the Ogoni Bill of Rights to the Federal Republic of Nigeria; and the Niger Delta Environmental Survey (NDES) was established to increase focus on Shell's operations in Nigeria. Internationally, in an endeavour to reduce oil spills which are caused due to oil piracy, the United States donated three 56 m (180 ft) refitted World War II-era patrol boats to the Nigerian Navy.[29] Additionally, several laws, both local and international, exist including among others the Oil Pollution Act (OPA) of 1990, Federal Environment Protection Agency Act Cap 131 LFN 1990, the Environmental Impact Assessment Decree No 86 of 1992, Convention on the Prevention of Marine Pollution Damage, 1972, and African Convention on the Conservation of Nature and Natural Resources, 1968, among others.

---

[28] See, UNEP, Environmental Assessment of Ogoniland Report, 2011, it can be accessed at https://postconflict.unep.ch/publications/OEA/UNEP_OEA.pdf

[29] For a full discussion on this, see Allison, C., Oriabure, G., Ndimele, P.E. and Shittu, J.A., 2018. Dealing with oil spill scenarios in the Niger Delta: Lessons from the past. In *The political ecology of oil and gas activities in the Nigerian aquatic ecosystem* (pp. 351–368).

Concluding Remarks on Fossil Fuels
Although oil spills and gas flaring are very important topics globally, which deserve much attention due to their contribution to climate change and their negative impact to the environment and human health, the focus in this chapter, however, is to highlight the fact that fossil fuels if well managed in a sustainable manner can contribute to tackling the challenge of energy access in SSA. As noted earlier, the flared gas in Nigeria is capable of meeting the energy needs of the country together with other neighbouring countries in West Africa. Relating to the main topic of this book regarding regionalism, West African countries through their regional organisation ECOWAS can strengthen regional cooperation in the utilisation of natural gas as a way of tackling the challenge of energy access in the region.

In terms of regional cooperation in the development of the oil and gas sector, there are regional efforts to establish infrastructure, as will be discussed in the next chapter. There are also plans in the SADC region to establish a Regional Petroleum and Gas Association (REPGA), aimed at promoting trade in the oil and gas sector. The Regional Petroleum and Gas Association is also anticipated to play a big role in the harmonisation of standards and regulations in the oil and gas sector; co-ordinate hydrocarbon exploration activities; and harmonise laws and regulations.[30]

4.3 Trilemma Between Climate Change, Energy Access, and Renewable Energy

Having established the challenges relating to the extraction of fossil fuels such as climate change, under this section the trilemma and nexus between climate change, energy access, and renewable energy will be discussed. To appreciate how renewables can play a role in tackling the challenge of climate change and energy poverty, we need to first understand the climate change impacts in SSA. In this regard, part one of this section sets out the panoramic overview of potential climate change impacts in SSA.

[30] SADC Energy Monitor, 2016. Page 77.

4.3.1 Panoramic Overview of Climate Change Impacts in SSA

Typically, climate change in simple terms can be understood to mean a change in the usual weather found in a place. According to the Oxford Dictionary, "climate change is a change in global or regional climate patterns, in particular a change apparent from the mid to late twentieth century onwards and attributed largely to the increased levels of atmospheric carbon dioxide produced by the use of fossil fuels".[31] This definition indeed brings into play the role of human activities and their contribution to climate change. The Intergovernmental Panel on Climate Change (IPCC), on the other hand, defines climate change as the state of climate that can be identified by changes in the mean and/or the variability by its properties and that persists for an extended period, typically decades or longer.[32] These changes affect the general environment and this in turn not only affects humans but also other species. It has been recognised that one of the most important challenges is the anticipation of the future under climate change; and indeed the concept has attracted a lot of attention in recent years from different experts and researchers of various backgrounds including scientists, oceanographers, ecologists, economists, and policy analysts.

Whereas the concept of climate change has gained a lot of attention in recent decades and as such it might be understood as a new challenge in recent years, it is worth noting that the extreme changes in the weather, which are now referred to as "climate change", have been in existence since time immemorial. Floods, droughts, and other extreme weather events have been experienced in various parts of the globe leading to not only food insecurity, displacement, and death but also the outbreak of dangerous diseases. For instance, in the period around 1626–1640 droughts were accompanied by regular floods in China, whereas in 1627 and 1630,

[31] See Oxford Online Dictionary, https://en.oxforddictionaries.com/definition/climate_change, last accessed on 22 November 2017.

[32] The International Panel on Climate Change (IPCC) 2007.

disastrous floods in Gujarak and Gujarat led to 3 million losing their lives. Moreover, it has been recognised that the mid-seventeenth century saw the weakest period of monsoons on record.[33]

The question of whether all the extreme weather changes that were experienced in the distant centuries were connected to human activity such as extraction of fossil fuels and coal is still open for discussion. Moreover, if the answer to that is positive, then a thorough reflection on why the international community did not act in those centuries might also be worth exploring. Could it be that in the distant decades and centuries many developed countries were relying on fossil fuels and coal for their energy needs and economic development, and thus ignored their negative impacts? Is it fair that now, after these countries have attained economic development and also overcome energy poverty, they deem it fit to react to climate change and also point out the negative effects of fossil fuels and coal? Or could it simply be that in the past there was no scientific proof that human activities influenced climate change?

Reflecting on these questions is indeed important given the recent threats by President Donald Trump of the United States to question the genuineness of the Paris Agreement, which Agreement represents the international response to climate change. These questions among others may well lie behind some scepticism about climate change in various countries including the United States and the United Kingdom, where it is believed that such scepticism has increased over the period 2006–2010.[34]

Researchers have gone ahead to distinguish such scepticism into three categories including trend sceptics (trend sceptics are those who deny there is such a thing as any upward trend in global temperatures); attribution sceptics (attribution sceptics are those who accept that the world's climate may be changing but do not think that this is

[33] See V. Damodaran and others, Climate Signals, Environment and livelihoods in C17th India in a comparative context. http://foodsecurity.exeter.ac.uk/wp-content/uploads/2015/09/17th-century.pdf

[34] Steentjes, K., Pidgeon, N.F., Poortinga, W., Corner, A.J., Arnold, A., Böhm, G., Mays, C., Poumadère, M., Ruddat, M., Scheer, D. and Sonnberger, M., 2017. *European Perceptions of Climate Change (EPCC): Topline findings of a survey conducted in four European countries in 2016.*

caused by human activity but rather by natural causes); and impact sceptics (impact sceptics are those who agree that the world's climate is changing as a result of human activity but do not think it will lead to substantial detrimental impacts).[35] The public perception about climate change can also explain the various scepticisms in various countries; for instance in a recent survey carried out in Europe, it was revealed that the people in the United Kingdom appeared to be the least worried about climate change with only 20% indicating that they were very or extremely worried about climate change and 38% saying they are not very worried or not worried at all.[36] Although the United Kingdom has a considerable number of people who don't believe in climate change, it is worth noting that the country presents itself as a champion of climate change, as envisaged with the enactment of the Climate Change Act of 2008—The Climate Change Act of 2008 among others mandates legally binding emission reduction of 80% by 2050 and the formation of the Committee on Climate Change to advise government on interim carbon budgets.

Despite the existence of scepticism, the need to make climate change an important challenge in the current global, regional, and national affairs was summarised by McNutt as follows:

> *The natural systems that provide oxygen, clean water, food, storm and erosion protection, natural products and the potential for future resources such as new genetic stocks for cultivating, must be protected, not because it is part of good stewardship but also so that they can take care of us.*[37]

The above statement clearly summarises the reason why climate change has become an important topic in recent years, as the failure to take good care of the natural systems has direct negative consequences to not only humans but also nature at large.

---

[35] V. Damodaran and others, Climate Signals, Environment and livelihoods in C17th India in a comparative context. http://foodsecurity.exeter.ac.uk/wp-content/uploads/2015/09/17th-century.pdf

[36] Ibid.

[37] McNutt, M., 2013. Climate change impacts. *Science (New York, NY), 341*(6145), p. 435.

4.3.2 Climate Change Impacts in SSA

It has been argued that the potential future effects associated with climate change can no longer be ignored, including more frequent wildfires, longer periods of drought, and an increase in the number, duration, and intensity of tropical storms.[38] Some of these have already been experienced in SSA and also in the developed world. Taking an example of Europe, it has been submitted that the Southern and Central Europe are seeing more frequent heat waves, forest fires, and droughts, while the Mediterranean area is becoming drier and Northern Europe is getting significantly wetter, making it vulnerable to winter floods.[39]

Being that the chapter and book generally focus on a single region, SSA, it is worth giving a detailed brief about the impacts of climate change on this segment of the African continent. Basically, before discussing the impacts of climate change in SSA, it is worth pointing out the general characteristics of the African climate, which is shaped by the inter-tropical convergence zone, seasonal monsoons in East and West Africa, and the EI Niño/La Nina Southern Oscillation (ENSO) in the South. The ENSO and the seasonal Monsoons influence temperatures and precipitation across the continent including extreme events in metrological droughts.[40]

Africa is the most vulnerable continent to climate change impacts, as it is expected to severely disrupt water and food systems, public health, and agricultural livelihoods not to mention causing enhanced droughts, sea level rise, and changes in the incidence and prevalence of vector-borne disease. These projected changes are expected to exacerbate already high levels of food and water insecurity, poverty, and poor health and undermine economic development.[41] In addition climate change impacts to

[38] NASA. Global Climate Change, https://climate.nasa.gov/effects/, last visited on 7 May 2017.

[39] European Commission: Climate Action, https://ec.europa.eu/clima/change/consequences_en, last visited on 8 May 2017.

[40] For a full discussion see Henderson, J.V., Storeygard, A. and Deichmann, U., 2017. Has climate change driven urbanization in Africa? *Journal of Development Economics, 124*, pp. 60–82.

[41] Adenle, A.A., Ford, J.D., Morton, J., Twomlow, S., Alverson, K., Cattaneo, A., Cervigni, R., Kurukulasuriya, P., Huq, S., Helfgott, A. and Ebinger, J.O., 2017. Managing climate change risks in Africa-A global perspective. *Ecological Economics, 141*, pp. 190–201.

the agricultural sector are likely to drive rapid urbanisation in Africa, as people will be pushed from rural areas to urban areas due to the deteriorating agricultural productivity caused by climate change.

There is a general view that Africa has witnessed warming over land consistent with anthropogenic climate change. This has indeed had enormous effects on the continent and it is predicted that these consequences will exacerbate both the economic and social situation on the continent in various ways such as affecting ecosystems, amplifying existing stress on water availability, worsening the existing health vulnerability, including insufficient access to safe water and improved sanitation.[42]

Taking the example of the Southern African region, the effects of climate change in the form of frequency of extreme weather events such as droughts and floods are not only evident in sectors such as agriculture and fishery but also present in the energy sector. This has therefore made it crucial for the region to ensure the deployment of climate resilient energy assets. A case in point is the El Niño climate event in Southern Africa, which has left an approximate of 21.3 million people in the region requiring emergency assistance due to the drought it has caused since 2015. Basically, El Niño is a naturally occurring phenomenon that involves fluctuations of sea surface temperatures and winds across the equatorial Pacific Ocean. Historically, it raises chances of receiving below average rainfall during the main crop growing season in Southern Africa.[43] This El Niño has deteriorated various sectors such as agriculture, food security, livestock, nutrition and water, and sanitation and hygiene conditions in countries such as Lesotho, Madagascar, Malawi, Mozambique, Swaziland, Zambia, and Zimbabwe.[44]

[42] Niang, I., O.C. Ruppel, M.A. Abdrabo, A. Essel, C. Lennard, J. Padgham, and P. Urquhart, 2014. Africa. In *Climate change 2014: Impacts, adaptation, and vulnerability. Part B: Regional aspects*. Contribution of Working Group II to the Fifth Assessment Report of the Intergovernmental Panel on Climate Change [Barros, V.R., C.B. Field, D.J. Dokken, M.D. Mastrandrea, K.J. Mach, T.E. Bilir, M. Chatterjee, K.L. Ebi, Y.O. Estrada, R.C. Genova, B. Girma, E.S. Kissel, A.N. Levy, S. MacCracken, P.R. Mastrandrea, and L.L. White (eds.)]. Cambridge University Press, Cambridge, United Kingdom and New York, NY, USA, pp. 1199–1265. Can be accessed at, http://www.ipcc.ch/pdf/assessment-report/ar5/wg2/WGIIAR5-Chap22_FINAL.pdf

[43] See, FEWS NET: Southern Africa Special Report, 2014/15 EL Niño Event, July 2014.

[44] USAID, Southern Africa Drought, Fact Sheet, Jan 2017. It can be accessed at https://scms.usaid.gov/sites/default/files/documents/1866/southern_africa_dr_fs04_01-30-2017.pdf, last visited on 8 May 2017.

With regard to the energy sector, countries that rely heavily on large hydropower schemes have indeed been affected with the climate change impacts such as droughts. A case in point is the SADC country of Zambia, which was left facing a 560 megawatt power deficit due to the reduced water levels at the Kariba lake reservoir. Indeed research on the effects of climate change on the Zambezi River Basin points to the fact that dry climate will typically reduce hydropower generation for both new and existing plants; as such it has been found necessary to not only seek other alternative energy but also to integrate both climate change and upstream development demands into the feasibility studies before investment decisions are made.[45]

In the Eastern Africa region, the impacts of climate change have also been evident in countries such as Kenya, South Sudan, and Uganda, which have been hit with a major drought thus leading to famine in various parts of these countries.

Reflecting on the discussion above, it goes without saying that Africa will experience diverse and severe impacts of climate change, making adaptation essential in these countries. Adaptation refers to the efforts across scales to build resilience and reduce vulnerability to the impacts of climate change. However, this faces many constraints in different African countries including among others insufficient climate data; limited engagement of adaptation responses to national planning processes and local expertise; failure to make adaptation responses broad so as to not only cover climate change but also climate variability and broader development issues; and insufficient adaptation finances. Additionally, there are other challenges facing adaptation in Africa including technical, political, institutional, economic, and social dimensions. For instance, with regard to the technical challenges, it is hard to develop better projections of climate change in African countries (although this is important for adaptation) and this is due to a lack of historical information on weather and climate.[46]

[45] Spalding-Fecher, R., Joyce, B. and Winkler, H., 2017. Climate change and hydropower in the Southern African Power Pool and Zambezi River Basin: System-wide impacts and policy implications. *Energy Policy, 103*, pp. 84–97.

[46] Adenle, A.A., Ford, J.D., Morton, J., Twomlow, S., Alverson, K., Cattaneo, A., Cervigni, R., Kurukulasuriya, P., Huq, S., Helfgott, A. and Ebinger, J.O., 2017. Managing climate change risks in Africa-A global perspective. *Ecological Economics, 141*, pp. 190–201.

As highlighted in the discussion above, climate change can also have adverse effects on the energy sector and this itself can slow down the ongoing efforts to tackle the challenge of energy access in SSA. In this regard, it is appropriate to briefly review the major regional and international responses to tackle climate change. This therefore takes us to the next discussion, regarding global responses to climate change.

4.3.2.1 Global and Regional Responses to Climate Change

The complexity of climate change lies in the way it impacts on or is impacted by global issues such as poverty, population growth, economic development, security, sustainable development, and resource management. This therefore makes climate change a "threat multiplier" which affects other sectors. It is on this basis that there has been an increase in policy responses to the issue of climate change as evidenced nationally, regionally, and internationally.

Although this book focuses on the regional responses, a brief analysis of the international response to climate change is appropriate given recent developments aimed at addressing the challenge, in addition to the close connection between energy access and climate change. Moreover, it is also essential for us to explore how international responses such as the Paris Agreement are supporting regional responses, specifically in analysing regional responses to energy access in SSA.

International Response to Climate Change

Although the issue of climate change was brought to the attention of the international community as far back as 1979, when the First World Climate Conference was held, no international convention or treaty was approved to address the issue in that period. It was not until 1992 that the UN Framework Convention on Climate Change (UNFCC) was adopted, and this marked the first international legal instrument to address the issue of climate change.[47] Moreover, after the adoption of the UNFCC, the international community experienced several advancements in line with climate change mitigation, including holding of the first Conference of the Parties (COP1) in 1995 in Berlin; the establishment of the UNFCC Secretariat in 1996 to support action under the Convention; the adoption of the Kyoto Protocol in 1997 at COP3; and the COP21 which was held in Paris paving the way for the adoption of the Paris Agreement.[48]

The existence of various international instruments which either directly or indirectly have an impact on climate change cannot be denied. However, in this chapter the emphasis will be on the newly adopted Paris Agreement. The question that needs to be addressed in this chapter is whether the Paris Agreement advocates for regional cooperation as a way of not only tackling climate change but also energy access. In responding to this question, an examination of the provisions of the Paris Agreement will be carried out, to determine how the Agreement supports regional cooperation in tackling climate change and energy access for the developing countries in SSA.

The 2015 Paris Agreement

The Paris Agreement which was signed in 2015 strives to address and respond to climate change, which it considers to be an urgent threat.[49] Though the Agreement refers to climate change as an urgent threat as evidenced by scientific knowledge, it is worth pointing out the fact that these threats to the environment and climate generally have been in existence since time immemorial. Perhaps the threats were never viewed as urgent due to the fact that in the distant decades/centuries the subject did

[47] Prior to the 1992 Convention, there were two other international environmental treaties that addressed climate change indirectly. The amended 1987 Montreal Protocol on Substances and the 1979 Geneva Convention on long-Range Trans-boundary Air Pollution and its Protocols regulate the emission of noxious gases some of which are precursors of greenhouse gas. For a discussion on this see http://www.ciesin.org/docs/iucc201/fs201.html, last visited on 5 May 2017.

[48] UN Climate Change, http://www.un.org/climatechange/towards-a-climate-agreement/, last visited on 5 May 2017.

[49] Article 2 of the Paris Agreement, 2015.

not receive a lot of media coverage as it is the case today.[50] The wide media coverage of the subject in recent years has triggered not only enormous literature on the subject but also huge international response as envisaged by the high participation of various countries in the adoption of the 2015 Paris Agreement. However, considering the economic status of many countries in SSA, most of which are relying on their natural resources such as fossil fuels to not only address energy access but also to eradicate poverty, one may wonder whether the international climate change responses will negatively affect the economic progress of SSA countries.

It is true that the Agreement does expressly recognise how climate change responses might have an impact on equitable access to sustainable development and eradication of poverty. In this regard, the Agreement throughout its provisions makes it clear that developed countries will provide financial support to developing countries in the implementation of the Agreement, especially with reference to mitigation and adaptation.[51] The Agreement also further points out that developed country Parties shall take a lead in mobilising climate finance.[52] These are indeed positive signals in recognition of the economic discrepancies between the developed countries, which have already benefited from the dirty energy sources such as coal and fossil fuels to attain massive economic development and address energy poverty, and the developing countries, which are still struggling to realise full economic development and also overcome energy poverty. There is no doubt that there is massive poverty in SSA and it is true that there is some potential for fossil fuels and coal to contribute in overcoming these challenges. The Agreement indeed recognises the poverty challenge, albeit it neglects to address how clean technology can be deployed and utilised to mitigate climate change effects and also to ensure that developing countries can also benefit from their fossil fuels to achieve economic development, as was the case for the developed world in distant centuries.

---

[50] See discussion in preceding section, on the various climatic change conditions that have been experienced as far as the sixteenth Century.

[51] See Article 9 (1) of the Paris Agreement, 2015.

[52] Article 9(2) of the Paris Agreement, 2015.

There are indeed many issues that need to be addressed regarding the Paris Agreement. However, the focus of this chapter is in respect to the impact of the Agreement in ensuring an advanced role of REC in tackling both climate change and energy access in SSA, through enhancing more regional cooperation. Before examining how the Agreement addresses the role of REC, it is worth analysing the main purpose behind the Paris Agreement, which is enshrined in Article 2 and it states as follows:

> *Article 2 (1); The Agreement aims at strengthening the global response to the threat of climate change, in the context of* ***sustainable development and efforts to reduce poverty***[53]*, including by;*
>
> (a) *Holding the increase in the global average below 2 degrees Celsius above pre-industrial levels and pursuing efforts to limit the temperature increase to 1.5 degrees Celsius above pre-industrial levels, recognizing that this would significantly reduce risks and impacts of climate change;*
>
> (b) *Increasing the ability to adapt to the adverse impacts of climate change and foster climate resilience and low greenhouse gas emissions development, in a manner that does not threaten food production; and*
>
> *2) This Agreement will be implemented to reflect equity and the principle of common but differentiated responsibilities and respective capabilities, in the light of different national circumstances.*[54]

The aspect of sustainability is clearly echoed in the Agreement as being central to the efforts to respond to climate change. Whereas the aspect of sustainability encourages the need to embrace renewable energy sources which are in this context preferred to fossil fuels, it becomes unclear how developing countries will successfully eradicate poverty if they must abandon their natural resources such as fossil fuels and coal.

Moreover, we have recently witnessed some developed countries such as the United States threatening to pull out of the Paris Agreement on the

[53] Emphasis put on this phrase is only for the analytical purposes in this book.

[54] Article 2 of the Paris Agreement, 2015.

premise that it is a bad deal for the economy of the country.[55] This action by President Trump reflects the aspect of nationalism, which contradicts the principles of regionalism and internationalism. However it will be unsurprising if other countries especially those in the developing world take the same decision in future—that is if they consider it fit to really utilise their resources to overcome energy access and also eradicate poverty.

The above notwithstanding, in reference to the role of REC in tackling energy access and climate change, which is the pivotal topic of this chapter, the Agreement does not seem to recognise how positively RECs can contribute to the global efforts to tackle climate change. Not much reference is made to the subject of RECs, except in Article 20, which is to the effect that "the Agreement shall be open for signature and subject to ratification, acceptance or approval by states and Regional Economic Integration organizations that are Parties to the Convention…"[56] The Agreement further states that, "any REC that becomes a Party to this Agreement without any of its Member States being a Party shall be bound by all the obligations under this Agreement…"[57]

Reflecting on the discussion above, it can be argued that the Agreement does not give much room or support for the role of RECs in tackling climate change. There is no doubt that in SSA, where most RECs are still undergoing economic integration and where these RECs mostly play the role of coordination, it would be next to impossible for these bodies to make decisions on behalf of their member states, as is the case envisaged in the Agreement.[58] This therefore implies that RECs don't necessarily have a big role to play in the enforcement and implementation of the provisions of the Paris Agreement.

---

[55] Recently President Donald Trump made his intentions of pulling out of the Paris Accord very clear. For a full discussion on this, see http://www.telegraph.co.uk/news/2017/06/01/trump-pull-paris-accord-seek-better-deal/, last visited on 24 June 2017.

[56] Article 20 (1) of the Paris Agreement, 2015.

[57] Article 20 (2) of the Paris Agreement, 2015.

[58] Article 25(2) of the Paris Agreement is to the effect that RECs have the power to vote in matters within their competence.

Regional Response to Climate Change

Nevertheless, though the Agreement does not say much about RECs, it is worth noting that RECs in different parts of the globe have taken initiatives to tackle climate change even before the Paris Agreement was signed in 2015. Although it was only until 1922 that the international community filled the legal gap in the climate change sphere, we note that some national and regional efforts towards this direction had already been explored. In the regional organisation of the EU for instance, in 1990, the Council of the European Communities (EC) adopted a policy that provides for stabilising the emission of carbon dioxide, the most significant greenhouse gas, at 1990 levels by the year 2000. In SSA on the other hand, RECs have been involved in the establishment of institutions aimed at tackling the challenge of climate change. Taking the example of SADC, they established the SADC Climate Service Centre (CSC) in 1990.[59] Whereas the Centre is concerned with providing operational, regional services for monitoring and predicting extremes in climate condition, it also provides training in climate prediction for personnel in the National Meteorological/Hydrological Services. Moreover, energy is one of the weather-sensitive sectors it reached out to.

4.4 Nexus Between Climate Change, Energy Access, and Renewables

Climate change, sustainable development, and energy access remain major central, global, and long-term challenges facing human kind today. Typically, the nexus between energy access, climate change, and renewable energy lies in the need for clean energy in the form of renewables to not only reduce carbon emissions but also accelerate energy supply in SSA. Energy demand is expected to increase in SSA and this definitely has the potential of increasing GHG emissions, considering the high

[59] Initially it was established as the Drought Monitoring Centre in 1990 and there are four of these established in Africa.

dependence on fossil fuels and coal for electricity generation in countries such as South Africa. Moreover, as far back as the 1990s, scholars associated with a view that a society seeking sustainable development must among others utilise energy resources which cause no environmental impact, otherwise, those which cause no emissions to the environment.[60] However in spite of these scholarly views in the 1990s, the reliance on coal and other fossil fuels went unquestioned in countries such as the United States and China, which definitely reached economic development partly due to the utilisation of these resources.

In recent years however, on top of identifying mitigation and adaptation as generic policy responses to climate changes, the utilisation of renewables with the intention of minimising greenhouse gas emissions has not only seen the escalation and increase in debates to shift to cleaner energy but also the commitment of countries such as China, which were initially sceptical about global climate measures before. These countries now show their commitment in Paris to an absolute cap on emissions subject to international measurements, reporting, and verification (MRV), thus committing to reduce energy and carbon intensity. However, there are some scholars who don't believe in climate change; for instance, S. Fred Singer, an atmospheric physicist, has together with a number of other scientists and scholars publicly opposed the findings of the UN's Intergovernmental Panel on Climate Change.[61]

In the energy sector, the nexus can be witnessed on the negative effects of climate change, which are visible on the existing energy infrastructure. These energy infrastructures are negatively impacted mainly due to severe storms and heat waves; this therefore lags behind the efforts to address the challenge of energy access in SSA. Moreover, effects of climate change such as droughts disrupt the flow of renewable energy resources such as hydropower, a main source of electricity in SSA.

---

[60] Dincer, I. and Rosen, M.A., 1998. A worldwide perspective on energy, environment and sustainable development. *International Journal of Energy Research, 22*(15), pp. 1305–1321.

[61] S. Fred Singer is the founder of the Science and Environment Policy Project (SEPP) and organisation that has actively been challenging the published findings of the UN's IPCC since the 1990s. In fact, SEPP established the Leipzig Declaration, a statement of dissent from the 1997 Kyoto Protocol, which has been signed by over one hundred scientists and meteorologists. For a full discussion on this see National Association of Scholars, https://www.nas.org/articles/Estimated_40_Percent_of_Scientists_Doubt_Manmade_Global_Warming, last visited on 4 October 2017.

While not much has been written regarding the nexus between energy access and climate change, on the other hand, literature has flourished discussing the nexus between climate change and energy security. In this regard, three potential mechanisms for linking energy security and climate change have been suggested by some scholars. These include: (1) climate change may create second-order effects that may exacerbate social instability and disrupt energy systems; (2) it may directly impact energy supply and/or systems; (3) it may influence energy security through the effects of climate-related polices.[62] The first mechanism basically relates to the impacts of climate change such as droughts and food insecurity which may escalate wars in fragile developing countries thus disrupting the energy system, especially where rebel groups target energy infrastructure.[63]

The brief overview above summarises the nexus between climate change, energy access, and renewable energy. It also briefly states the nexus between energy security and climate change. The issue to be addressed in the next section is how the deployment of renewable energy can help tackle both climate change and energy access in SSA.

4.5 Renewable Energy Development as a Response to Climate Change and Energy Access in SSA

In this section, renewable energy developments will be discussed in respect to regional cooperation as a way of tackling the challenge of energy access in SSA. Prior to discussing regional policy and the institutional framework in place, a panoramic overview of RE development will be provided.

---

[62] King, M.D. and Gulledge, J., 2014. Climate change and energy security: An analysis of policy research. *Climatic Change, 123*(1), pp. 57–68.

[63] DoD, U., 2011. *Energy for the warfighter: Operational energy strategy*. Washington, DC: DD Deputy Secretary of Defence.

4.5.1 Panoramic Overview of Renewable Energy Development

The development of renewable energy sources in recent years has been central in not only addressing climate change but also tackling the challenge of energy access in developing countries, where the potential of renewables is now significant for both cheap and sustainable energy. Though in the developed world, the discussion on energy security mostly focused on maintaining access to fossil fuels, in recent years there has been a shift in the way energy security is viewed in these countries. In this regard, renewables are now also considered as a source of energy security. This shift indeed points to the flexibility of renewables to address the trilemma of energy access, energy security, and climate change. Moreover, some of the world's largest power producers such as China, who initially relied on massive fossil fuels and coal, are now leading in renewable energy systems, which are indeed expanding at a very high rate.[64] However, these countries still recognise and heavily rely on dirty energy such as fossil fuels and coal, which are central in meeting their energy needs.

Nevertheless, in an effort to mitigate the effects of fossil fuels and coal, endeavours have been made to deploy clean technology such as hydraulic fracturing combined with horizontal drilling for shale gas and carbon capture and storage (CCS) for coal. Moreover, the need to reduce carbon emissions has not only emphasised the role of renewable energy and the deployment of clean technology, but it has also triggered scholars to consider other mechanisms such as improved electricity storage as ways of curbing emissions, albeit this depends on the competitiveness of renewable energy against conventional electricity generation.[65]

Typically, by definition, renewable energies are energy sources that are continually replenished by nature and derived directly from the sun (such

---

[64] China, for instance, now leads the world in the production and use of wind turbines, solar-photovoltaic cells, and smart grid technologies. In 2013, the country had a total capacity of 378GW of renewable power. For a full discussion of this, see Mathews, J.A. and Tan, H., 2014. Manufacture renewables to build energy security. *Nature, 513*(7517), p. 166.

[65] For a detailed discussion on electricity storage, see Lazkano, I., Nøstbakken, L. and Pelli, M., 2017. From fossil fuels to renewables: The role of electricity storage. *European Economic Review*.

as thermal, photo-chemical, and photo-electric), indirectly from the sun (such as wind, hydropower, and photosynthetic energy stored in biomass), or from other natural movements and mechanisms of the environment (such as geothermal and tidal energy).[66]

4.5.2 Sources of Renewable Energy

The global effort and campaign of reducing carbon emissions have seen developed countries such as Spain achieve 100% electrification from renewables. This is indeed a goal shared by many developing countries in SSA, as evidenced in their pledge in 2016 during the United Nations Climate Change Conference in Morocco. Though the SSA region still faces the challenge of energy poverty, there is potential to meet this challenge by utilising the enormous renewable resources in the region. There is no doubt that SSA is very rich in renewable energy resources, with solar potential totalling about 10,000 GW; wind potential, totalling about 109 GW, mostly in the coastal countries; geothermal capacity estimated at 15 GW especially in the East African Rift Valley; and exploitable hydropower estimated at about 350 GW mainly located in Angola, Cameroon, Ethiopia, Gabon, and DRC.[67]

Despite these enormous resources and the global commitment to increase the percentage of renewables in the energy mix, there are some basic requirements that need to be fulfilled if the vision is to be attained. These requirements were summarised by Arila et al. in their renewable guide to include among others:

- policies that incentivise renewable energy deployment and *discourage new fossil fuel development*[68]

[66] Ellabban, O., Abu-Rub, H. and Blaabjerg, F., 2014. Renewable energy resources: Current status, future prospects and their enabling technology. *Renewable and Sustainable Energy Reviews, 39*, pp. 748–764.

[67] Avila, N., Carvallo, J.P., Shaw, B. and Kammen, D.M., 2017. *The energy challenge in sub-Saharan Africa: A guide for advocates and policy makers*. Technical report, OXFAM.

[68] Emphasis is for the purposes of this chapter.

- innovative financing mechanisms that allow decentralised solutions to thrive and be integrated into future grid expansion
- an enabling framework that attracts private investors to the energy sector and builds human capacity by empowering local entrepreneurs
- improved organisational procedures and sector management that support operational power pools powered by international and interregional transmission lines, allowing power sharing and reduced costs
- policies that incentivise and support decentralised renewable energy systems that enable 100% access to affordable electricity
- electricity supply strategies that prioritise the diversity of resources such as dispatchable renewables and storage negate the need for fossil fuels and ensure supply security[69]

Although some of the requirements outlined above are essential, others are impractical especially the first one relating to discouraging new fossil fuel developments. As highlighted in the previous sections, it is imperative to invest in clean technology instead of discouraging the deployment of fossil fuels.

Indeed, one of the prerequisites for utilising renewable energy resources is to ensure that the country and region adopt favourable renewable energy policies, competent enough to achieve the primary objective of maintaining or expanding energy services, improving energy access and security, and reducing health and environmental impacts of energy use.

Renewable energy sources basically include biomass, geothermal energy, hydropower energy, marine energy, solar energy, and wind energy. These are briefly discussed below:

1. *Biomass*

   Biomass is a source of energy through plants, crops, and trees and it entails the collection and storage of the sun's energy through photosynthesis. The biomass process also involves the conversion of biomass into useful forms of energy such as heat, electricity, and liquid fuels.

[69] Avila, N., Carvallo, J.P., Shaw, B. and Kammen, D.M., 2017. *The energy challenge in sub-Saharan Africa: A guide for advocates and policy makers*. Technical report, OXFAM.

2. *Geothermal energy*
   Geothermal energy is an efficient way to extract renewable energy from the earth, and it originates from the formation of the planet and from radioactive decay of materials.
3. *Hydropower energy*
   Hydropower is power derived from the energy of moving water. It is generated from water moving in the hydrological cycle, which is driven by solar radiation. Dams are the most prevalent form of hydropower.
4. *Marine energy*
   Marine energy, sometimes referred to as ocean energy, comes from six distinct sources, including waves, tidal range, tidal currents, ocean currents, ocean thermal energy conversion, and salinity gradients, each with different origins and requiring different technologies for conversion.[70]
5. *Solar energy*
   Solar energy is produced through sunlight. Solar energy generation involves the use of the sun's energy to provide hot water via solar thermal systems or electricity via solar photovoltaic (PV) and concentrating solar power (CSP) systems. Solar energy is mostly used for lighting purposes through photovoltaic conversion in the rural areas.
6. *Wind energy*
   Wind energy is produced by harnessing wind power. It is basically understood as the conversion of wind energy by wind turbines into a useful form, such as using wind turbines to make electricity, wind mills for mechanical power, wind pumps for pumping water or drainage, or sails to propel ships.

4.5.3 Advantages and Challenges of Renewable Energy

There are various advantages of renewable energy sources, for instance, hydro resources have considerable potential to be utilised for power

[70] Due to the absence of cost-effective technology, this type of energy has not yet been fully tapped in various African countries.

generation. On the other hand, solar and wind energy resources are considered to be excellent for applications such as water pumping, water heating, and power generation through centralised schemes, mini-grids, and stand-alone systems.[71]

Notwithstanding the advantages associated with renewable energy sources, there are some shortcomings relating to the reliance of renewables to expand the supply of electricity, which are prone to impacts of climate change hence hampering hydropower, the intermittency and variability of solar and wind, and the risk of over-generation and curtailment.[72] Moreover, some countries have already experienced the negative impacts of renewables such as overcapacity. A case in point is Spain, which has in recent years been experiencing overcapacity challenges, which have in turn blocked the deployment of renewable energy.[73] Spain's current situation dates back as far back as the 1990s when it started adopting effective support schemes for renewable energy in the power sector. This support from government increased the investments in RES thus contributing to the overcapacity situation, which necessitated the government of Spain in 2010 to make profound energy policy changes in the direction of reducing support for RES in order to convert the "tariff deficit" affecting the Spanish power system. Consequently, several legal actions have been brought against the Spanish government by investors in solar PV and other renewable energy projects as a result of regulatory changes which had the effect of cutting payments to renewable energy projects.[74]

The Spanish experience therefore highlights some of the challenges of renewable energy development in European countries, specifically challenges relating to overcapacity mainly caused due to inadequate

---

[71] Ershad, A.M., 2017. Institutional and policy assessment of renewable energy sector in Afghanistan. *Journal of Renewable Energy, 2017,* 1–10.

[72] Avila, N., Carvallo, J.P., Shaw, B. and Kammen, D.M., 2017. *The energy challenge in sub-Saharan Africa: A guide for advocates and policy makers*. Technical report, OXFAM.

[73] Del Río, P. and Janeiro, L., 2016. Overcapacity as a barrier to renewable energy deployment: The Spanish case. *Journal of Energy, 2016.*

[74] There are over 25 cases brought by international investors pending in the International Centre for Settlement of Investment Disputes (ICSID) against the Spanish government as a result of reductions in payments of renewable energy projects.

Table 4.1 Advantages and disadvantages of renewables[a]

Resource	Advantages	Disadvantages
1. Biomass	• Abundant and renewable • Reduce the global warming potential (waste based bio-refinery)	• Burning biomass can result to air pollution • May not be cost effective
2. Geothermal	• Unlimited supply of energy • Produces no air or water pollution	• Costly to establish and maintain
3. Marine	• Ideal for an island country • Captures energy that would otherwise not be collected	• Construction can be costly • Has negative impact to wildlife • Takes up lots of space
4. Solar/wind	• Potentially infinite energy supply • Causes no air or water pollution	• May not be cost effective • Storage and backup are necessary • Variable and intermittent (reliability depends on sunshine)
5. Hydropower	• Abundant, clean and safe • Easily stored in reservoirs • Relatively inexpensive way to produce electricity • Offers recreational benefits like boating and fishing	• Can cause flooding • Can threaten livelihood[b] • Pose social and ecological risks, especially large dams • Affected by drought and lack of water • May negatively impact wildlife

[a]Extracted from, Ellabban, O., Abu-Rub, H. and Blaabjerg, F., 2014. Renewable energy resources: Current status, future prospects and their enabling technology. *Renewable and Sustainable Energy Reviews*, 39, pp. 748–764
[b]Dams can threaten livelihood by causing the loss of agricultural land and requiring resettlement

planning for the future demand and supply. This is indeed an issue that SSA countries have to look into while initiating and implementing RES policies.

Additionally, the table below gives a general summary of both the benefits and barriers of each type of renewable energy sources (see Table 4.1):

Reflecting on the discussion above, it can be argued that although renewable energy is now central in tackling climate change, energy access, and energy security, the disadvantages and challenges associated with the development of renewable energy sources pose a significant barrier to their development. It is true that there are various key issues relating to renewables which are worth exploring. However, this chapter specifically focuses on regional perspectives of renewable energy development as a way of addressing the dilemma of climate change and energy access in SSA. Precisely, regional cooperation in the development of renewable energy will be discussed. In this regard, a three-step framework in the form of three sub-sections will be employed to present a general overview of the development of renewable energy together with regional policy and institutional frameworks in place to further develop renewable energy in SSA. Prior to discussing the institutional framework, it is worth discussing the impact of energy efficiency as this goes hand in hand with renewable energies as a way of responding to climate change and energy consumption.

4.6 Energy Efficiency in SSA

A discussion on renewable energy brings into play the issue of energy efficiency. As illustrated in the preceding sections, the global energy consumption is on the increase in many countries, leading not only to increased local air pollution but also greenhouse gas emissions. Energy efficiency and various technological advancements in the energy sector are considered as some of the available options for the reduction of carbon emissions. Moreover, energy efficiency is also considered as a complement of renewable energy—considering that the reduction of energy demand through energy efficiency is capable of improving the financial feasibility of renewable energy options.[75]

By definition, energy efficiency refers to the use of less energy to achieve the same services.[76] Although this definition clearly brings out the essence of energy efficiency, it has however been expanded by the IEA to bring out the multiplier benefits of energy efficiency including economic development, environmental sustainability, and enhanced

[75] SADC: Renewable Energy and Energy Efficiency Status Report, 2015. Page 11.

[76] See, International Energy Agency 2015 report. It can be accessed at https://www.iea.org/publications/freepublications/publication/Multiple_Benefits_of_Energy_Efficiency.pdf, last visited on 20 January 2018.

development.[77] Additionally, it is worth noting that consuming energy efficiently can also lead to reductions in the price of energy services and reduce demand, hence delaying the requirement for new generation capacity. For instance, demand can be reduced by replacing incandescent bulbs with compact fluorescent lamps (CFLs).

Energy efficiency has not only been advocated in household use, but also in the industry sector, which is believed to emit the largest share of GHG, (particularly the energy intensive industries such as chemicals, iron and steel, minerals, and petroleum refining).[78] It is against this background that energy policy discussions have increasingly focused on enhancing energy efficiency, as a mechanism of reducing emissions of carbon dioxide (CO_2) and other pollutants. Moreover, we cannot ignore the fact that energy efficiency is a cost-effective means of mitigating greenhouse gas emissions from industries, transport, cooking, and space heating just to mention but a few.

The increasing focus on energy efficiency is also due to the fact that most countries lack national policies to cap or tax CO_2 emissions, thus turning to the promotion of low or zero emitting energy technologies through a mixture of standards and incentives, as a way of responding to global warming.[79] With regard to the employment of energy efficient technologies in developed and developing countries, it has been highlighted that developed countries, given their ready access to advanced technology, tend to use less energy in their industrial sectors, as compared to developing countries, which are characterised by inefficient technologies to achieve energy efficiency and emissions reductions in their industrial sectors.[80] However, the challenge relating to the energy efficiency gap

---

[77] Ibid.

[78] Worrell, E., Bernstein, L., Roy, J., Price, L. and Harnisch, J., 2009. Industrial energy efficiency and climate change mitigation. *Energy Efficiency*, *2*(2), p. 109.

[79] Gillingham, K. and Palmer, K., 2014. Bridging the energy efficiency gap: Policy insights from economic theory and empirical evidence. *Review of Environmental Economics and Policy*, *8*(1), pp. 18–38.

[80] Worrell, E., Bernstein, L., Roy, J., Price, L. and Harnisch, J., 2009. Industrial energy efficiency and climate change mitigation. *Energy Efficiency*, *2*(2), p. 109. There is recent research on the subject including: Pierson, J.M. and Hlavacs, H., 2015. Introduction to energy efficiency in large-scale distributed systems. *Large-scale distributed systems and energy efficiency: A holistic view*, pp. 1–16; Gellings, C.W. and Parmenter, K.E., 2016. Energy efficiency in fertilizer production and use.

cannot be ignored in both the developed and developing countries (the energy efficiency gap is broadly described to refer to the slower than socially optimal rate of diffusion of energy-efficient products).[81] Although there have been several policies advocating energy efficiency, it is imperative to note that investments in energy efficiency are considered to be risky due to the fluctuating energy prices and the irreversibility of the investment. Additionally, other factors limiting the use of energy in an efficient manner include market failures (these basically include environmental and national security externalities, inefficient pricing of energy, and lack of information); behavioural explanations (these include inattentiveness and salience issues and systematically biased beliefs); and modelling flaws (these refer to reasons why the observed rate of diffusion of energy saving technology may be as paradoxical as it may at first appear).[82] There are indeed various issues associated with energy efficiency, but the focus in this book is to highlight how energy efficiency can contribute to ensuring energy access in SSA.

4.6.1 Energy Efficiency in SSA

Although there are common challenges faced by both the developed and developing countries as regards the enhancement of energy efficiency, there are some challenges which are unique to SSA countries. These include lack of local trained workforce, poor regulatory environment and governance, and lack of access to financing for energy efficiency projects.

Notwithstanding the challenges encountered in employing energy efficient techniques, many governments and firms around the world have adopted policies and programmes to capture the benefits that accrue from energy efficiency. In SSA, several countries have employed different energy efficiency activities including basic CFL replacement programmes

[81] Gillingham, K. and Palmer, K., 2014. Bridging the energy efficiency gap: Policy insights from economic theory and empirical evidence. *Review of Environmental Economics and Policy, 8*(1), pp. 18–38.

[82] See Gerarden, T.D., Newell, R.G. and Stavins, R.N., 2017. Assessing the energy-efficiency gap. *Journal of Economic Literature, 55*(4), pp. 1486–1525.

(in the SADC region for instance, over ten member states have instituted CFL replacement programmes)[83]; solar water heating; demand market participation; standards and labelling; hot water load control; awareness programmes; and energy audits in the industrial and building sectors. In Ghana for instance, they introduced a programme for labelling appliances, aimed at revealing to the consumers the energy consumption and efficiency of the product. These efforts together with the regulatory framework have led to an estimated peak energy savings of over 120 megawatts (MW). Additionally, the programme has saved the country USD 105 million in generation investment and reducing carbon dioxide emissions by over 11,000 tons annually.[84]

In South Africa, the establishment of energy efficiency incentive programmes by Eskom led to the saving of over three gigawatts (GW) of total cumulative energy.[85] In Namibia, in an endeavour to increase research in energy efficiency, the Department of Civil Engineering built a demonstration and research house, where 60–70 energy is saved in the residential building through thermal envelope, air tightness, and sub-soil heat exchangers.[86] Mauritius on the other hand instituted a National Energy Efficiency Programme. Improved cook stove programmes are also being embraced in various SSA countries as a form of energy efficiency: In the SADC region for instance, all the countries except Mauritius and Seychelles have programmes aimed at increasing the use of energy efficient cook stoves.[87]

In terms of regional efforts, various institutions including the SAPP in SADC have played a big role in promoting energy efficiency including developing specific programmes for CFL replacement and initiating an expanded Energy Efficiency Framework: The Framework covers four technologies including CLFs, Commercial lighting retrofits, solar water

[83] See SADC: Renewable Energy and Energy Efficiency Status Report, 2015. Page 11.

[84] See USAID: Examining energy efficiency issues in Sub-Saharan Africa, https://www.usaid.gov/powerafrica/newsletter/dec2014/smarter-power-in-africa, last accessed on 20 January 2018.

[85] Ibid.

[86] SADC REESAP 2016–2030, Renewable Energy and Energy Efficiency Strategy and Action Plan. Page 42.

[87] SADC: Renewable Energy and Energy Efficiency Status Report, 2015. Page 11.

heating, and distribution transformer retrofits. Additionally, in the SADC region, there are more than five national utilities which, on the basis of the SAPP initiative, developed demand-side management (DSM) on their own.[88]

In terms of energy intensity (this is used as a measure of energy efficiency of an economy),[89] some countries are doing better than others, for instance in the SADC region, the DRC, Mozambique, and Zimbabwe have the highest, at 19.1, 17.9, and 17.5 MJ, respectively.[90] Due to the employment of energy efficiency in the SADC region, there was a demand energy reduction of 4500 MW by the end of 2015.[91]

4.7 Regional Cooperation in the Development of Renewables

The global effort to employ renewable energy has witnessed numerous countries around the globe setting renewable energy targets and polices. By early 2014, at least 144 countries had renewable energy targets and 138 countries had renewable energy support policies in place.[92] In a regional perspective, there have been efforts to cooperate in the regional development of renewable energy as a way of tackling both climate change and the challenge of energy access in SSA. In this regard, there have been various regional renewable energy policies set in SSA, and these have their genesis from the various establishing treaties of the RECs as will be discussed in the next section.

---

[88] Ibid.

[89] Energy intensity is calculated as units of energy per unit of Gross Domestic Product (GDP), and this implies that if the energy intensity is high, then the cost of production in the country is also high. Energy intensity is usually measured in mega joules (MJ) per USD of GDP. See the Energy Monitor 2016. Page 71.

[90] Energy intensity varies across regions and countries, for instance, in the SADC region it is at an average of 9.4 MJ per USD of GDP, while in ECOWAS it is estimated at 14.5 MJ. See SADC: Renewable Energy and Energy Efficiency Status Report, 2015. Page 11.

[91] See the SADC Energy Monitor 2016. Page 71.

[92] REN21, Renewables 2014: Global Status Report, REN21 Bonn, Germany 2014.

4.7.1 Regional Legal Frameworks

4.7.1.1 Regional Treaties

Typically, treaties are the legal basis for regional cooperation in the development of the different energy sectors including the renewable energy sector. The SADC Treaty for instance under Chap. 7 provides for the different areas of regional cooperation, amongst which is cooperation in infrastructure and environment.[93] The Treaty does not specifically refer to cooperation in the energy sector, more so to the renewable energy sector. It is worth noting that the SADC Treaty does not expressly make a mention of regional cooperation in the energy sector and specifically the renewable energy sector. Nevertheless, this falls under infrastructure, which is expressly mentioned in the Treaty.

Comparatively, the ECOWAS Treaty under Chapter V expressly mentions the need for cooperation in the energy sector[94] and environment.[95] Whereas the ECOWAS Treaty goes ahead to mention the energy sector, it does not make a specific reference to renewables.

Whereas the establishing Treaties do not make a descriptive reference to how cooperation in the energy sector will be conducted, these Treaties nevertheless provide for the establishment of protocols which provide for the various areas of cooperation in the energy sector. The SADC Treaty for instance under Article 22 provides that "*member states shall conclude such Protocols as may be necessary in each area of cooperation, which shall spell out the objectives and scope of, and institutional mechanism for cooperation and integration.*"[96] The ECOWAS Treaty mentions the need for a regional energy policy; it states under Article 28 that "*member states shall articulate a common energy policy, particularly in the field of research exploitation, production and distribution*".[97] These Treaties indeed are the basis for regional cooperation in the development of renewable energy in

[93] Article 21 (3) (a) (b) of the SADC Treaty.

[94] Article 28 of the ECOWAS Treaty.

[95] Article 29 of the ECOWAS Treaty.

[96] Article 22 (1) of the SADC Treaty.

[97] Article 28 (2) (e) of the ECOWAS Treaty.

SSA. The next section examines the energy protocols and their effect on the development of the renewable energy sector in SSA regional organisations.

4.7.1.2 Regional Energy Protocols

Typically, regional energy protocols are legal documents which stipulate in detail how countries will cooperate in the energy sector. In SSA, these have been established in various RECs including SADC and ECOWAS. However, as will be discussed in detail, some of these protocols are either outdated or do not expressly and technically address current energy challenges relating to renewable energy and climate change.

Analysing the SADC Energy Protocol for instance, we note that this was signed by the member states in 1996 at the time when renewable energy and climate change were not focal topics. The Protocol has some essential underlying principles such as the need to develop and use energy in an environmentally sound manner;[98] the need to increase private sector participation; and the necessity for member states to fully accept the responsibility of sharing the costs associated with institutional mechanisms created for the effective implementation of the Protocol.[99] The Protocol is undoubtedly outdated and as such does not elaborate in detail how renewable energy and climate change can be addressed regionally. The Protocol nevertheless recognises energy cooperation in the SADC region as its main objective,[100] and indeed it mentions regional cooperation in the development of new and renewable energy sources, energy efficiency, and conservation.[101] Some of the issues not addressed in the SADC Protocol include principles regarding competition in the energy sector, principles and rules governing trade amongst members, and principles and rules regarding establishment of refineries in the SADC region.

---

[98] Article 2 (8) of the SADC Protocol on Energy.

[99] See Article 2 (5) of the SADC Protocol on Energy.

[100] Article 3 (1) of the SADC Protocol on Energy.

[101] Article 3 (1) of the SADC Protocol specifically mentions regional cooperation in the development of new and renewable energy sources, energy efficiency, and conservation.

Moreover, there is no provision or legislation on how the private companies in the renewable energy sector such as solar energy companies interact or coordinate with the public. This principle is necessary in SSA because there are increased investments in the renewable energy sector and other low-carbon technologies such as electric vehicles (EVs). For instance, the IEA records show that in the year 2016, more than 750,000 electric cars were sold worldwide, taking the global stock to more than 2 million.[102]

With regard to investments in renewables, during my fieldwork research at the East African Community Secretariat in Arusha, Tanzania, I had the chance of visiting some foreign companies involved in the business of solar panels and other renewable energy equipment. One issue that stood out especially with relation to energy access is the fact that companies such as Mobisol had an instalment payment programme for their customers (most of which are low-income earners).[103] Whereas this instalment payment method was favourable to enable low-income earners access solar energy, in the event the client failed to pay the whole amount in an agreed time, then all the solar equipment would be retained by the company. This method as it stands cannot permanently solve the challenge of energy access in Africa. It is just a temporary solution which definitely emphasises the need for African countries through the various regional organisations to regulate and supervise the activities of various energy investors, specifically with a view of protecting consumers.

Moving forward, it is worth noting that whereas the SADC Energy Protocol is outdated and lacks some principles, comparatively, the ECOWAS Protocol on energy is elaborative and covers a wide range of topics and issues which are neither addressed in EAC or SADC. Some of these issues include among others principles relating to competition in the energy sector (here contracting parties are obligated to alleviate market distortions and barriers to competition in the energy sector)[104]; principles and rules on transit (the Protocol makes it mandatory for

[102] International Energy Agency: World Energy Outlook, 2017.

[103] Mobisol operates in three East African countries including Rwanda, Kenya, and Tanzania. See the company official website at http://plugintheworld.com/about/

[104] Article 6 (1) of the ECOWAS Energy Protocol. The provision further emphasises the need for contracting parties so that such competition laws are enforced within national jurisdiction; see Article 6(2) of the ECOWAS Energy Protocol.

contracting parties to ensure the facilitation of transit of energy materials and products)[105]; the Protocol further regulates on the promotion and protection of energy investors in the areas of the contracting parties,[106] together with principles of transfer of technology[107] and compensation for losses.[108] Although the ECOWAS Protocol on Energy is elaborative on certain principles as discussed here, it is imperative to note that the Protocol does not give much attention to the aspect of renewable energy and how this can be developed regionally to tackle the challenge of both climate change and energy access. Article 43 of the Protocol covers environmental aspects and Article 43 on the other hand covers issues relating to energy efficiency; however, these are not elaborative enough to cover current challenges relating to energy access and climate change.

4.7.1.3 Other Regional Instruments

Besides the various RECs Treaties and energy protocols, there are other instruments that have an impact on not only the renewable energy sector but energy generally. These take the form of master plans and other regional programmes. In SADC for instance, the Energy Sector Plan (ESP), which is under the auspices of the SADC Regional Infrastructure Development Master Plan, is intended to address four key strategic objectives including ensuring energy security, improving access to modern energy services, tapping the abundant energy resources, and achieving financial investment and environmental sustainability. One of the sectors covered by the ESP includes renewable energy and energy efficiency.[109] Furthermore, in SADC, besides the master plan there have been an implementation of programmes in the renewable energy sector in line with the SADC Energy Protocol including the following: the Energy Sector Plan of the SADC Regional Infrastructure Development Master

[105] See Article 7 of the ECOWAS Energy Protocol.

[106] Article 10 of the ECOWAS Energy Protocol.

[107] Article 8 of the ECOWAS Energy Protocol.

[108] Article 12 of the ECOWAS Energy Protocol.

[109] Other sectors covered in the ESP are electricity, coal, petroleum, and gas.

Plan (REASAP, 2012); the Regional Energy Access Strategy and Action Plan (REASAP, 2012); the Renewable Energy and Energy Efficiency Strategy & Action Plan (REEESAP 2016–2030); the Programme for Biomass Energy Conservation (ProBEC); and the United Nations Development Programme-supported Financing Energy Services for Small-Scale Energy Users Project.[110]

In ECOWAS, besides the Energy Protocol and the Treaty, the ECOWAS/UEMOA White Paper on access to energy services for populations in rural and peri-urban areas was adopted in 2006, and this encourages the use of renewable energy in reaching the electrification goals.[111]

4.7.2 Regional Institutional Response

Besides the establishment of various laws and policies, the efforts to mainstream renewable energy and energy efficiency (RE & EE) have been experienced in SSA at the regional level through the establishment of regional centres. There is no doubt that SSA REC through their various activities in the renewable energy sector aim at meeting the objectives of the UN's Sustainable Energy for All initiatives. The establishment of RE & EE regional centres indeed follows the successful establishment of regional power pools such as SAPP and EAPP in Southern and Eastern Africa, respectively. Whereas regional power pools are mostly concerned with power trading, the RE & EE centres are mostly concerned with the promotion of RE & EE technologies and the development of markets. This is envisaged through sharing information and best practices; developing sound policy, regulatory, and legal frameworks; and building the capacity within the member states of RECs concerned.

These centres are at different stages of development with some RECs such as ECOWAS having functional institutions and others such as EAC and SADC being in the preparatory stages of establishing these institutions. In West Africa, the ECOWAS Regional Centre for

[110] SADC, Energy Monitor, 2016. Page 55.

[111] ECOWAS Centre for Renewable Energy and Energy Efficiency, http://www.ecreee.org/page/overview-ecreee, last visited on 5 June 2017.

Renewable Energy and Energy Efficiency (ECREE) legally came into existence by the adoption of Regulation C/REG 23/11/08 in 2008 at the 61st Session of ECOWAS Council of Ministers—and the secretariat of the centre was established in Praia, Cape Verde in 2010.[112]

Although we note that in West Africa the renewable centre has been in existence for more than a decade, in East and Southern Africa, preparations are still underway to establish the regional renewable energy centres. In EAC for instance, the East African Centre for Renewable Energy and Energy Efficiency (EACREE) was approved during the 33rd Meeting of Council of Ministers held on 29 February 2016. In fact, Makerere University College of Engineering, Art, Design and Technology (CEDAT) was designated as a Centre of Excellence for EACREE. In Southern Africa, the establishment of the SADC Centre for Renewable Energy and Energy Efficiency (SACREE) was approved by the SADC energy ministers on 24 July 2015.

Whereas the objectives of RE & EE centres are promising, it is imperative to note that these will not be achieved by the mere establishment of these centres. There is a need to strengthen not only regional institutions such as regional regulator associations, but also to establish and strengthen national institutions.[113] These are necessary to adopt and implement regional RE & EE projects. However, we note that not all countries are at the same level of establishing the necessary institutions. In SADC for instance, the Regional Electricity Regulatory Association (RERA) is comprised of only 12 Regulatory Agencies implying that three SADC member states have not yet set up national regulatory authorities. Moreover, in order to achieve regional renewable energy targets, there is a need to establish renewable energy agencies and national frameworks for RE in all member states of various RECs.[114]

---

[112] ECOWAS Centre for Renewable Energy and Energy Efficiency, http://www.ecreee.org/page/overview-ecreee, last visited on 5 June 2017.

[113] SADC Renewable Energy and Energy Efficiency Strategy and Action Plan (REESAP 2016–2030). Page 35.

[114] Ibid.

Due to the various challenges in different countries such as political instability, lack of technical expertise, and financial constraints, SSA RECs are facing discrepancies in the development of RE policies and frameworks at the national level. Taking the example of SADC, South Africa seems to be a step ahead when it comes to the national policies of RE. For instance in 2011 the Department of Energy (DoE) launched the Renewable Energy Independent Power Producer Procurement Program (REIPPP or REI4P) and this is used to tender large-scale installation including technologies such as solar PV, onshore wind, small hydro, landfill gas, and biomass.[115]

There are therefore various challenges when it comes to implementing both regional policies and institutional mechanisms aimed at promoting the development of renewable energy as a mechanism of tackling energy access and climate change in SSA.

4.8 Conclusion

Regional cooperative development of renewable energy can contribute to addressing the challenge of energy access in SSA. However, as discussed in the previous sections SSA's regional organisations are not only struggling with the challenge of inadequate finances but they also have weak legal and institutional frameworks focusing on renewable energy. This therefore affects the ongoing initiatives to tackle both climate change and energy access in SSA. Additionally, the selfish interest of countries to reach economic development and overcome energy access and energy security challenges also presents a situation where global responses to climate change will be undermined, especially if countries continue to heavily rely on dirty energy sources such as fossil fuels.

Whereas it is acceptable to use renewable energy resources to overcome the challenge of climate change and energy access, it is imperative to note that this should not be done at the expense of other energy resources such as fossil fuels and coal. History agrees with the notion that many developed

[115] Ibid. Page 36.

countries such as those in Europe, the United States, and Asia have efficiently utilised fossil fuels and coal to achieve the level of the development they are at now. This should be a case for the developing world by embracing clean technology to utilise those resources.

In this regard and following the discussion in this chapter, below are the proposals and recommendations to be considered:

Given the crucial role of renewables and following the fast development in this sector, there is a need for SSA regional organisations to strengthen the established renewable energy centres in these regions. The legal mandate and jurisdiction of the regional renewable centres should expand to from merely publishing reports to making decisions in the administration and establishment of coordinated renewable energy projects in the region. These centres could also be utilised in the areas of investment and these could also be the basis for improving regulation of renewable energy and harmonisation of such regulation at the regional level. Additionally, the global responses to climate change such as the Paris Agreement should consider getting closely involved in regional organisations so as to utilise the already existing regional cooperative mechanisms to tackle both climate change and energy access through renewable energy.

References

Adenle, A. A., Ford, J. D., Morton, J., Twomlow, S., Alverson, K., Cattaneo, A., et al. (2017). Managing climate change risks in Africa-A global perspective. *Ecological Economics, 141*, 190–201.

Akinola, A. O. (2017). Resource mis governance and the contradictions of gas flaring in Nigeria: A theoretical conversation. *Journal of Asian and African Studies*. https://doi.org/10.1177/0021909617722374.

Allison, C., Oriabure, G., Ndimele, P. E., & Shittu, J. A. (2018). Dealing with oil spill scenarios in the Niger Delta: Lessons from the past. In *The political ecology of oil and gas activities in the Nigerian aquatic ecosystem* (pp. 351–368).

Avila, N., Carvallo, J. P., Shaw, B., & Kammen, D. M. (2017). *The energy challenge in sub-Saharan Africa: A guide for advocates and policy makers*. Technical report, OXFAM.

Del Río, P., & Janeiro, L. (2016). Overcapacity as a barrier to renewable energy deployment: The Spanish case. *Journal of Energy, 2016*.

Dincer, I., & Rosen, M. A. (1998). A worldwide perspective on energy, environment and sustainable development. *International Journal of Energy Research, 22*(15), 1305–1321.

ECOWAS Energy Protocol.

ECOWAS Treaty.

Ellabban, O., Abu-Rub, H., & Blaabjerg, F. (2014). Renewable energy resources: Current status, future prospects and their enabling technology. *Renewable and Sustainable Energy Reviews, 39*, 748–764.

Ershad, A. M. (2017). Institutional and policy assessment of renewable energy sector in Afghanistan. *Journal of Renewable Energy, 2017*, 1–10.

Gerarden, T. D., Newell, R. G., & Stavins, R. N. (2017). Assessing the energy-efficiency gap. *Journal of Economic Literature, 55*(4), 1486–1525.

Gillingham, K., & Palmer, K. (2014). Bridging the energy efficiency gap: Policy insights from economic theory and empirical evidence. *Review of Environmental Economics and Policy, 8*(1), 18–38.

Hassan, A., & Kouhy, R. (2013, June). Gas flaring in Nigeria: Analysis of changes in its consequent carbon emission and reporting. In *Accounting Forum* (Vol. 37, No. 2, pp. 124–134). Elsevier.

Henderson, J. V., Storeygard, A., & Deichmann, U. (2017). Has climate change driven urbanization in Africa? *Journal of Development Economics, 124*, 60–82.

International Energy Agency: World Energy Outlook, 2017.

King, M. D., & Gulledge, J. (2014). Climate change and energy security: An analysis of policy research. *Climatic Change, 123*(1), 57–68.

Lazkano, I., Nøstbakken, L., & Pelli, M. (2017). From fossil fuels to renewables: The role of electricity storage. *European Economic Review.*

McNutt, M. (2013). Climate change impacts. *Science (New York, NY), 341*(6145), 435.

Niang, I., Ruppel, O. C., Abdrabo, M. A., Essel, A., Lennard, C., Padgham, J., et al. (2014). Africa. In *Climate change 2014: Impacts, adaptation, and vulnerability. Part B: Regional aspects.* Contribution of Working Group II to the Fifth Assessment Report of the Intergovernmental Panel on Climate Change [Barros, V. R., Field, C. B., Dokken, D. J., Mastrandrea, M. D., Mach, K. J., Bilir, T. E., Chatterjee, M., Ebi, K. L., Estrada, Y. O., Genova, R. C., Girma, B., Kissel, E. S., Levy, A. N., MacCracken, S., Mastrandrea, P. R., & White, L. L. (Eds.)] (pp. 1199–1265). Cambridge, UK and New York: Cambridge University Press.

Paris Agreement, 2015.

SADC REESAP 2016-2030, Renewable Energy and Energy Efficiency Strategy and Action Plan, Page 42.

SADC: Renewable Energy and Energy Efficiency Status Report, 2015. Page, 11.
SADC Protocol on Energy.
SADC Renewable Energy and Energy Efficiency Strategy and Action Plan (REESAP 2016-2030). Page 35.
SADC Treaty.
SADC, Energy Monitor, 2016. Page 55.
Spalding-Fecher, R., Joyce, B., & Winkler, H. (2017). Climate change and hydropower in the Southern African Power Pool and Zambezi River Basin: System-wide impacts and policy implications. *Energy Policy, 103*, 84–97.
Steentjes, K., Pidgeon, N. F., Poortinga, W., Corner, A. J., Arnold, A., Böhm, G., et al. (2017). *European Perceptions of Climate Change (EPCC): Topline findings of a survey conducted in four European countries in 2016.*
The International Panel on Climate Change (IPCC) 2007.
UNEP, Environmental Assessment of Ogoniland Report, 2011, it can be accessed at https://postconflict.unep.ch/publications/OEA/UNEP_OEA.pdf
V. Damodaran and others, Climate Signals, Environment and livelihoods in C17th India in a comparative context.
Worrell, E., Bernstein, L., Roy, J., Price, L., & Harnisch, J. (2009). Industrial energy efficiency and climate change mitigation. *Energy Efficiency, 2*(2), 109.
Zabbey, N., Sam, K., & Onyebuchi, A. T. (2017). Remediation of contaminated lands in the Niger Delta, Nigeria. *Science of the Total Environment, 586*, 952–965.

5

Regional Cooperation in the Establishment of Regional Energy Infrastructure

5.1 Introduction

Energy resources are unequally distributed across the globe and this has necessitated the transnational character in the extraction and utilisation of these resources. This makes the establishment and development of polycentric regional energy infrastructure crucial, as it enables both the producing and consuming countries to benefit from the various energy resources.[1] Although fundamental for tackling the challenge of energy access, the establishment and development of energy infrastructure generally raises a lot of concerns both nationally and internationally—for instance where issues arise concerning consumer privacy[2]; discontent by local people leading to protests over the combined environmental and economic harm caused by energy. Regional challenges arise where there are complex issues regarding legal framework and financial contribution of the partner states. Moreover, political challenges including geopolitics,

[1] Goldthau, A., 2014. Rethinking the governance of energy infrastructure: Scale, decentralization and polycentrism. *Energy Research & Social Science, 1*, pp. 134–140.

[2] See Quinn, E.L., 2009. Privacy and the new energy infrastructure, https://papers.ssrn.com/sol3/papers.cfm?abstract_id=1370731, last accessed on 14 February 2017.

V. R. Nalule, *Energy Poverty and Access Challenges in Sub-Saharan Africa*, Energy, Climate and the Environment, https://doi.org/10.1007/978-3-319-95402-8_5

lack of political will, and insecurity are also a hindrance to the realisation of regional energy projects including regional energy infrastructure.

Critical energy infrastructure includes utilities associated with energy transport and management such as electricity generation and transmission; oil and gas production and pipeline systems; liquefied natural gas facilities; coal transport trains; management technology such as advanced electricity metering and distribution systems; and modern power plant control systems and smart building technologies. Several of these have been established not only in SSA, but also other parts of the globe like Europe and America. However, looking at SSA, a number of additional issues arise when considering whether regional energy infrastructure developments enhance regional cooperation as a mechanism of tackling the challenge of energy access in SSA: Are the existing regional cooperative mechanisms in the establishment and development of regional energy infrastructure in SSA both legally and institutionally effective to address the challenge of energy access? Are there other alternatives that can be employed in the region to advance regional energy infrastructure development?

In addressing the issues above, regional organisations and specifically the RECs will be discussed as a mechanism of regional cooperation. The focus will be limited to an examination of EAC, ECOWAS, and SADC. RECs have become popular in SSA and these are used as a basis for promoting greater economic integration. Whereas RECs play a big role in the establishment of regional energy infrastructural projects, in most cases they lack the technical and logistical expertise, thus necessitating support from specialised institutions such as power pools and river basin commissions in implementing infrastructure policies and projects. With this in mind, this chapter examines how different regional institutions such as RECs and specialised institutions such as regional power pools—as mechanisms of regional cooperation—enhance the establishment and development of regional energy infrastructure in tackling the challenge of energy access in SSA.

Electricity interconnections as a form of regional energy infrastructure will be the focus of this chapter. Interconnections have been embraced and encouraged in different parts of the world including in Europe for instance, where the creation of a common market and integrated market for electricity is the major goal of the European Commission (EC). This is envisaged in the EU Directive 2009/72/EC, which placed wider emphasis on cross-border interconnections and the need to mitigate barriers to cross-border trade.[3] These electricity interconnections have been associated with increasing competition, improving market integration and power trading. Additionally, some of the advantages of larger and integrated electricity markets include enhanced security of supply and a reduction in reserve capacity needed to maintain a given level of system performance.

Although the chapter concentrates on electricity interconnections, oil and gas pipelines are also worth discussing given their contribution in addressing energy security concerns: In West Africa for instance, the establishment of the West African Gas Pipeline (WAPP) is considered a major step in tackling the challenge of energy access in the region. Both pipelines and electricity interconnection grids are essential elements in ensuring energy trade. However, this dichotomy of exporting energy when shortfalls exist in the domestic market has been cited as one of the biggest domestic political constraints to energy cooperation in developing regions.[4] Despite some shortfalls, the benefits associated with regional energy infrastructure are considered to be substantial, including lower investment requirements; better coordination; reduction in system operational costs; supply of reliable and affordable electricity, and so on.[5] However, operation of regional infrastructure remains very complex, involving multi-disciplinary issues, which need to be addressed through cooperation among the countries concerned. It is imperative though to bear in mind that RECs are multinational bodies, and whereas they lack the financial ability to sponsor regional energy infrastructure, their role is still essential considering the sovereign and nationalistic (sometimes selfish) needs of different countries who might give less attention to regional

[3] For a detailed discussion on this, see Nepal, R. and Jamasb, T., 2012. Interconnections and market integration in the Irish Single Electricity Market. *Energy Policy, 51*, pp. 425–434.

[4] Huda, M.S. and McDonald, M., 2016. Regional cooperation on energy in South Asia: Unraveling the political challenges in implementing transnational pipelines and electricity grids. *Energy Policy, 98*, pp. 73–83.

[5] Armar, A., 2009. *Building regional power pools. A toolkit* (p. 2). Washington, DC: The World Bank; Jerome, A. and Nabena, D., 2016. Infrastructure and regional integration in Africa. In *Region-building in Africa* (pp. 89–108). Palgrave Macmillan US.

initiatives such as regional energy infrastructure developments. In this regard therefore, the role of RECs becomes more visible.

In addressing the questions raised, the chapter employs a three-step framework to examine the issues. First, a panoramic overview of the general regional energy infrastructure developments is presented. This section is essential as it illustrates the regional developments in tackling the challenge of energy poverty through regional energy infrastructure. Moreover, the section helps us to highlight the influence of different regional institutions in the establishment and development of regional energy infrastructure.

In the second section, RECs with specific focus on EAC and ECOWAS are examined through a comparative analysis to highlight their influence in the establishment and development of international power grids as a form of regional energy infrastructure. This section examines the institutional and legal framework of RECs and how these contribute to the establishment and development of regional energy infrastructure. The section also explores the legal basis for RECs involvement in regional energy matters and examines the specific energy laws and policies in place that have been developed by EAC and ECOWAS. The last section provides concluding remarks and briefly discusses specialised regional institutions such as power pools, specifically emphasising their role in the establishment and development of power grids as a form of regional energy infrastructure. Although it is true that power pools are mostly concerned with power trading, given their growing role in the development and establishment of soft energy infrastructure, it is imperative to discuss them as a mechanism of regional cooperation in the establishment and development of regional energy infrastructure.

Moreover, financial institutions such as the World Bank, AFDB, and development partners are now dispatching infrastructural finances through RECs and specialised institutions such as regional power pools. This therefore enhances the role of power pools and other specialised institutions in the establishment and development of regional energy infrastructure in SSA. This chapter will therefore consider the influence of not only RECs but also specialised institutions in the establishment and development of regional energy infrastructure as a mechanism to deal with the issue of energy access in SSA.

5.2 Regional Energy Infrastructure Establishment

The development of regional energy infrastructure has been evident in different parts of the globe. Though this chapter concentrates mostly on on-grid electricity interconnection, it is important to note the rise in off-grid electricity interconnections such as solar and wind energy, which are now being embraced in developing countries in an endeavour to tackle the challenge of energy access. Moreover, oil and gas pipelines have also attracted a lot of attention mostly due to the rising hydrocarbon discoveries in different parts of SSA including Western and Eastern Africa. In this section, oil and gas pipeline developments and electricity interconnections will be discussed as a form of energy infrastructure.

5.3 Establishment of Oil and Gas Pipelines

Oil and gas pipelines are considered to have the highest capacity and to be the safest and least environmentally disruptive means for oil or gas transmission. These have been established in different regions; in Europe for instance, over 27 oil and gas pipelines exist. The development of oil and gas pipelines historically dates as far back as the 1850s, when the first commercial oil well was drilled in Pennsylvania in 1859, necessitating the construction of the first cross-country pipeline in 1879.[6] To date, oil and gas pipelines are considered to be a very important form of energy infrastructure considering the vast amount of oil and gas consumed around the world. Pipelines transport oil, natural gas, and petroleum products from producing fields and refineries to consumers, sometimes crossing other countries on a transit basis. Moreover, in instances where the discoveries are offshore, pipelines are used to gather the resources and bring them onshore.

The extraction of oil and gas together with the development of pipelines attracts several issues among which we may include conflict with the

[6] See John L. Kennedy, 1993. *Oil and gas pipeline fundamentals* (p. 2). Pennwell Books.

indigenous people's territories and environmental damages. Pipelines can also lead to hazardous consequences and large business losses if not well maintained through repairs, rehabilitations, or replacements. Moreover, technical difficulties, financial constraints, bureaucratic inefficiency, and political challenges have also been cited as major constraints to the development of these pipelines in different regions.[7] This notwithstanding, oil and gas pipelines are generally commended for their remarkable efficiency and low transportation cost and reduced damage to the environment considering the kind of technology employed.

Although there are various issues associated with oil and gas pipelines, such as those relating to failures of the pipelines due to corrosion defects, which are considered a major concern in pipeline integrity,[8] this section will focus the discussion to specifically examine the role of RECs in fostering regional cooperation in the development of oil and gas pipelines. A comparative analysis will be employed to evaluate how different RECs in SSA influence the development of oil and gas pipelines.

5.3.1 Oil and Gas Pipelines in SSA

Oil and gas pipelines have been developed in different parts of SSA to enhance the utilisation of fossil fuels. Looking at West Africa, under the auspices of ECOWAS, the West African Gas Pipeline (WAGP) was established with the main purpose of constructing a 600 km pipeline to transport natural gas from Nigeria to Benin, Togo, and Ghana, the four main countries in the project. Prior to discussing WAGP, it is worth noting that the West African region is very rich in hydrocarbon resources. However, over 200 million people do not have access to electricity or mechanical power and as such are dependent on traditional forms of energy.[9]

---

[7] Huda, M.S. and McDonald, M., 2016. Regional cooperation on energy in South Asia: Unraveling the political challenges in implementing transnational pipelines and electricity grids. *Energy Policy, 98*, pp. 73–83.

[8] For a detailed discussion on this, see Choi, J.B., Goo, B.K., Kim, J.C., Kim, Y.J. and Kim, W.S., 2003. Development of limit load solutions for corroded gas pipelines. *International Journal of Pressure Vessels and Piping, 80*(2), pp. 121–128.

[9] See ECREE: ECOWAS Energy Efficiency Policy (EEEP), September, 2012. Page 8. http://www.ecreee.org/sites/default/files/documents/basic_page/081012-ecowas-ee-policy-final-en.pdf

In an attempt to tackle the energy challenges in the region, WAGP was established based on the Treaty of the West African Gas Pipeline Project.[10] ECOWAS as a regional organisation is not directly involved in the operation of WAPP, rather the pipeline is owned and operated by the West African Gas Pipeline Company Limited (WAPCo), with a regulatory body based in Abuja, the West African Gas Pipeline Authority. The pipeline is anticipated to reduce overall production costs of electricity generation using fossil fuels; improve energy access; and reduce gas flaring which is associated with environmental pollution.

The example of RECs such as ECOWAS as a mechanism of regional cooperation in tackling energy access in SSA is illustrated especially by drawing on their reliance of specialised institutions such as WAGP in the establishment and development of regional oil and gas pipelines. Another important observation in such regional efforts is the fact that it is mostly the State that is directly concerned with the infrastructure, and it is they who take the lead role: although the project is under the auspices of ECOWAS, the four States (Benin, Ghana, Nigeria, and Togo) are directly involved in WAGP. Moreover, a specialised institution under the auspices of the WAGP Authority oversees the operations of the WAGP, thus limiting the role of ECOWAS as an REC.[11]

One can therefore argue that the establishment and development of regional energy infrastructure is an individual effort of the countries intended to benefit from such a project. As such, the aspect of regionalism becomes less operative if the project concerns only a few countries in the region. In the case of WAGP, only four countries out of the 15 member states of ECOWAS have a lead role. Moreover, ECOWAS as an REC also has limited influence on the project.

The limited influence of regional bodies in energy infrastructure projects is due to the financial nature of energy projects, which necessitate billions of monies. In this regard, host countries often engage more with international investors to finance energy infrastructure such as oil and gas

[10] The WAGP Treaty was signed on 31 January 2002 between Benin, Ghana, Nigeria and Togo.

[11] The role of ECOWAS in managing such energy infrastructure may be limited due to the rise of these specialised institutions but it is worth noting that ECOWAS still carries out its role as a facilitator and coordinator of such projects. This will be further illustrated in the next section while discussing the influence of RECs in regional energy infrastructural developments.

pipelines. This therefore brings into play the aspect of individualism of States thus weakening the role of regional bodies, which in most cases lack the financial capacity to develop such energy infrastructure.

Whereas the influence of ECOWAS in West Africa in the establishment and development of oil and gas pipelines is more visible through specialised institutions, in East Africa by comparison, the EAC has not yet employed a specialised institution. Rather the EAC organs take part in the decision-making process concerning the relevant regional energy infrastructure development. Nonetheless, the individual States concerned mostly take the lead role, although the EAC Secretariat takes up the role of facilitator and coordinator of such energy projects until project commissioning when the partner states take control over the project. Examples of regional oil and gas projects in the EAC are discussed below.

In the EAC, taking the example of Uganda (a landlocked country with proven oil reserves of 6.5 billion barrels, but dependent on its neighbour for oil transportation), it is in the process of constructing the Uganda-Tanzania Crude Oil Pipeline (UTCOP) to transport crude oil from Ugandan fields to Tanga, Tanzania, a port on the Indian Ocean. Previously, Uganda had considered partnering with Kenya to construct an oil pipeline (the Uganda-Kenya Crude Oil Pipeline); however, this idea was abandoned due to security and cost concerns, making Uganda to opt for a route in Tanzania. Another project which is attracting regional efforts is the construction of an oil refinery in western Uganda, which is aimed at processing oil for exports within the EAC. This project is to be developed under a public private partnership and the EAC countries are expected to have a stake in this USD 2.5 billion project. Tanzania for instance agreed to buy 8% of the 40 shares in the refinery allocated to EAC countries for USD 150.4 million.[12]

Given other various oil and gas infrastructure anticipated to be developed in the region, such as the Kigali-Bujumbura Oil Products Pipeline Project and the Mbarara-Mwaza-Ishaka-Dar es Salaam Oil Products Pipeline Project, it is worth investigating if EAC also needs a specialised institution in the establishment and development of oil and gas pipelines as is the case in ECOWAS with the establishment of WAGP.

---

[12] See Kasumuni, Ludger (30 April 2016). TZ to spend Sh300bn on buying stake in Uganda refinery. *The Citizen (Tanzania)*. Dar Es Salaam. Retrieved 5 March 2017.

If the role of regional organisations in addressing the challenge of energy access is to be more effective, there is a need to enhance regionalism in the establishment and development of oil and gas infrastructure in SSA as opposed to individualism. In this regard, and drawing from the discussion above, EAC should consider establishing a specialised institution mandated to coordinate the establishment and development of oil and gas infrastructure in the region just like ECOWAS. Additionally, ECOWAS should make its specialised institution more inclusive by increasing the level of participation from its member states. Alternatively, given the fact that RECs are involved with various sectors such as agriculture, health, trade, and many others, which of course require special attention and are still lacking in many SSA countries, and reflecting on the discussion in Chap. 2 of this book, RECs should be thinking of establishing a more focused regional organisation in SSA concerned specifically with tackling energy access. This organisation could also be responsible for ensuring the establishment and development of the necessary oil and gas infrastructure as a form of addressing the challenge of energy access.

The above responses are proposals and recommendations that RECs could consider to more effectively tackle energy access. In summary, EAC needs a specialised institution to handle energy infrastructure establishment and developments in the region. ECOWAS, in contrast, must make its institution more engaging. These developments are essential to ensure effective regional cooperation in tackling the challenge of energy access in SSA. Having discussed oil and gas infrastructure, the next section examines electricity interconnections as a form of regional energy infrastructure. This section will contribute to addressing the questions raised in this chapter regarding the role of RECs in the establishment and development of regional energy infrastructure in SSA.

5.4 Power Grid Interconnections

International power grids have been embraced all over the globe as a form of ensuring energy security and in SSA in particular they are essential in tackling the challenge of energy access. They have become indispensable for modern economies to secure electricity supply; moreover, energy security is now one of the most important policies in terms of economic activities. There have been visible electricity access challenges in various African countries; taking the example of the SADC region, some countries opted for load-shedding as a stop-gap measure in order to conserve energy and these include among others Botswana, Namibia, South Africa, Zambia, and Zimbabwe. Due to energy shortages in the SADC region, in January 2008, South Africa had to temporarily shut down its mining industry: one which accounts for 15% of the country's electricity demand.[13] There are some reasons identified for the electricity deficit in the SADC region: an increase in the population of most SADC member states; economic expansion necessitating new industries; limited capital injection into electricity generation projects; rise in power demand; aged power generation assets, and so on.[14] All these challenges make regional energy infrastructure such as electric power grids establishment and development inevitable.

By definition, the electric power grid is understood as the entire apparatus of wires and machines that connects the sources of electricity, the power plants, with customers and their myriad needs.[15] In other words, it connects supply with demand. International power grids can take different forms depending on the number of countries involved and the amount of electricity being transferred. They can be as modest as the one-way transfer of a small amount of electricity from one country to another or as ambitious as the full integration of the power systems and markets of all the countries in the region.[16] Once loosely interconnected networks of largely local systems, electric power grids increasingly host large-scale, long-distance transmission of power from one region to another.[17]

[13] See SADC Energy Monitor, 2016. Page 32.

[14] Ibid.

[15] Amin, M., 2008, July. Challenges in reliability, security, efficiency, and resilience of energy infrastructure: Toward smart self-healing electric power grid. In *Power and energy society general meeting-conversion and delivery of electrical energy in the 21st century, 2008 IEEE* (pp. 1–5). IEEE.

[16] Hector G Sarmiento, Miguel Angel Avila: Issues Associated with International Power Grid Interconnections in Mexico.

[17] Amin, M., 2008, July. Challenges in reliability, security, efficiency, and resilience of energy infrastructure: Toward smart self-healing electric power grid. In *Power and energy society general meeting-conversion and delivery of electrical energy in the 21st century, 2008 IEEE* (pp. 1–5). IEEE.

These interconnections are associated with several advantages: creating incentives for new investments; meeting the growing demand for electric energy; ensuring efficient operation of generation; exchange of reserve (generation) capacity among national power utilities; enabling the retirement of old and polluting generating plants; enhancement of stability and quality of supply; increase in capacity energy interchange and sale opportunities; and provision for back-up transmission in a particular area or country.[18] An interconnected system is also economically justifiable because it incurs lower operating cost by permitting excess supply in one node to be utilised in other nodes where the marginal cost would be higher if there were no interconnection. Moreover, these could also tackle environmental concerns. From an environmental point of view, interconnection gives incentives to substitute cleaner fuels for coal and oil in more competitive electricity markets.

Notwithstanding the associated advantages outlined above, the aspect of electricity interconnection, be it with two countries or more, brings into play the notion of regional energy cooperation, which of course comes with various complexities in legal, social, and economic spheres. One of such complexities is the need for deregulation of the electricity sector for successful interconnection. In the next section, electricity interconnections in SSA will be discussed, specifically looking at the EAC and the ECOWAS regions.

5.4.1 Power Grid Interconnections in SSA

The growing need for modern economies to secure electricity supply has necessitated the establishment and development of power grid interconnections around the globe. In recent years, there have been initiatives and efforts to develop an energy system that can utilise hydro resources and other types of energy sources on a regional, continental, and intercontinental basis. Although there are several advantages for regional and continental efforts in the utilisation of energy resources, it is imperative to

[18] Hector G Sarmiento, Miguel Angel Avila: Issues Associated with International Power Grid Interconnections in Mexico.

note that this can also be a disadvantage especially due to the fact that it can increase the sensitivities to dependency on other states for energy.

As far back as 1997, there were suggestions on the establishment and implementation of an integrated African grid, which would serve as a technically reliable and economic foundation for stimulating economic growth throughout the continent.[19] Arguments raised for an integrated African grid, include an enhanced system operational reliability and flexibility as a result of creating adequate spinning reserve and improved voltage regulation; establishment of an integrated water management programme and therefore an improved flood control, transportation, and the resulting social and environmental benefits; coordinated infrastructural development such as access to roads and communications to support access to and operation of the central grid; satisfying system and structural developmental programmes aspired by the African Union (AU); creating an atmosphere of enhanced awareness and synergy among African policy makers by the globalisation of local African conditions; and creating an atmosphere of enhanced awareness and synergy among African engineers and international research institutions and policy makers by the globalisation of local African conditions.[20]

Though the ideas of an integrated African grid are good, making the idea a reality may be difficult. First and foremost, we have to consider the disparities in development in the different African countries some of which are still plagued with political insecurities. Nevertheless, Blyden made some recommendations on how to implement the integrated African grid. He recognised the need to internationalise the African grid initiative by consolidating relevant energy, economic, and environmental data resources through a dedicated web site.[21] The integrated African grid can however be a reality if the regional grids are effective.

In this section, we examine how regional bodies in SSA are involved in these interconnections, and the issue which arises here is whether this practice enhances the role of regional cooperation in tackling the challenge

[19] Blyden, B.K., 1997. Towards the implementation of an integrated African grid. *IEEE Power Engineering Review, 17*(1), pp. 11–11.

[20] Blyden, B.K., 1997. Towards the implementation of an integrated African grid. *IEEE Power Engineering Review, 17*(1), pp. 11–11.

[21] Ibid.

of energy access. In answering this, a comparative analysis looking into electricity interconnections in the Eastern and Western Africa will be employed to highlight the role of EAC and ECOWAS in the establishment and development of electricity infrastructure in the region.

In various RECs, the energy sector is one of the most important sectors in the industrial and economic integration of these economic organisations, be it in SSA or other parts of the globe. Taking the example of ECOWAS, the region embraced power grid interconnections as a mechanism of energy cooperation as far back as the 1960s when the West African Grid System (WAGS) was established.[22] The genesis of WAGS can be traced to 1969 when the Volta River Authority (VRA) of Ghana initiated its interconnection project with the Communauté Électricité du Benin (CEB) of Togo and Benin. Further expansion of the WAGS included an interconnection with Cote d'Ivoire.

Among the numerous important issues that arise as a result of regional electricity interconnection, there is an issue regarding management of the established infrastructure by the countries concerned. Other issues include the necessary legal framework, financial contribution, and the social and environmental impacts of such infrastructure, to mention but a few. This aspect of management requires effective cooperative mechanisms in both the establishment and maintenance of the project in question. Looking at WAGS, there are separate technical committees that have been formed between VRA and CEB, and between VRA and CIE to take decisions on matters concerned with the technical management of each tie-line. These committees approve all operating instructions and technical studies undertaken on the tie-line.[23] Additionally, they prepare protocols for energy exchanges and undertake negotiations on matters concerning power sales on behalf of the management of their respective utilities.[24]

However, whereas in ECOWAS there is a common practice of forming specialised institutions such as WAGS (for electricity interconnections)

[22] Ghana's experience in power interconnection within the framework of ECOWAS, http://repository.uneca.org/bitstream/handle/123456789/17587/Bib-67662.pdf?sequence=1, last visited on 20 March 2017.

[23] Ibid.

[24] Ibid.

and WAGP (for oil and gas pipelines), this practice has not yet been embraced by EAC, except in the field of renewable energy and power pooling. Prior to discussing the involvement of RECs in regional energy infrastructure, the different regional interconnectors in EAC will be highlighted.

5.4.1.1 Examples of Regional Electricity Interconnection Projects in EAC

Electricity interconnection projects among EAC partner states are some of the efforts to electrify the East African region. The table below highlights some of the interconnection projects in the EAC region (see Table 5.1).

The table above clearly illustrates why regional cooperation is important in the establishment and development of energy infrastructure such as electricity interconnection networks. All EAC partner states have ongoing electricity interconnection projects amongst themselves. Additionally, non-partner states such as Ethiopia and the DRC are also involved in planned interconnections in EAC. This situation puts EAC in a pivotal role but as will be discussed in the next section, looking at some of the challenges in RECs such as the lack of capacity at EAC (both human and financial capacity), specialised institutions such as the East African Power Pool tend to be more suitable to handle energy infrastructure projects.

Referring to the table above again, whereas some projects (1, 3, and 8) are still undergoing construction, other projects (7, 9, and 10) have not yet identified the financiers.[25] The EAC Secretariat in most cases is responsible for the sourcing of funding for feasibility studies to be carried out regarding the various energy infrastructural projects in the region. This brings into play the relationship between regional bodies and international financial institutions such as the World Bank and AFDB: together with development partners such as the US Agency for International Development (USAID), the UK Department for International Development (DFID), and the German Society for

[25] 11th Meeting of the Sectorial Council on Energy Report of the Meeting of the Energy Committee, EAC Headquarters, Arusha, Tanzania, 14–16 March 2016. Page 10.

Table 5.1 EAC interconnection projects[a]

Name	Description
1. Kenya-Uganda Interconnector	132 km, 400 kV double circuit line and upgrading
2. Kenya-Tanzania Interconnector	507 km 400 kV double circuit line between Isinya and Namanga
3. Ethiopia-Kenya Interconnector	612 km 500 kV HVDC bipolar line from Suswa to the Ethiopia-Kenya boarder, converters and 400 kV substation at Suswa
4. Rwanda-Burundi Interconnector	Kigoma-Gitega OHTL, 220 kV 143 km and associated substation works
5. Masaka-Mwanza Interconnector	Construction of 220 kV, 335 km, Masaka-Mwanza transmission line and associated substations
6. Iringa-Shinyanga via Singida backbone transmission line	Construction of 400 kV, Iringa-Shinyanga via Singida transmission line investment project
7. Iringa-Mbeya-Tunduma backbone transmission line	Construction of 392 km, 400 kV Transmission line and Iringa-Mbeya-Tunduma
8. Uganda-Rwanda Interconnector	Construction of 66 km, 220 kV transmission line: Mbarara-Mirama-Birembo
9. Uganda-DRC	Construction of 72.5 km, 220 kV transmission line: Nkenda-Mpondwe (Uganda part)
10. Uganda-South Sudan	Construction of 190 km, 400 kV transmission line: Olwiyo-Nimule (Uganda part)
11. Rwanda-Burundi Interconnector	Kigoma-Gitega OHTL, 220 kV 143 km and associated substation works
12. Burundi-DRC	Kamanyola-Bujumbura OHTL 220 kV and associated substation Bujumbura-Uvira OHTL 110 kV and associated substation

[a]Extracted from the Report of the 11th Meeting of the Sectorial Council on Energy, Report of the Meeting of the Energy Committee, EAC Headquarters, Arusha, Tanzania, and 14—16 March 2016.

International Cooperation (GIZ) among others. Energy projects are very expensive ventures, and as such the role of international financial institutions and their relationship with regional bodies is very important in the establishment and development of energy infrastructure in many developing regions such as SSA, as these can be useful in financing some of the feasibility studies for energy projects.

Besides power grid interconnections, there are other power projects in EAC: Kikagati/Murongo Hydropower Projects, between the United

Republic of Tanzania and the Republic of Uganda[26]; and the Nsogenzi 35 MW hydropower project, again between Uganda and Tanzania.[27] Although hydropower projects are also essential in addressing energy poverty in SSA, the lack of financing capability and the reluctance of international organisations to be involved in such projects are some of the cited major obstacles to hydropower facilities; this was the case with the Souapiti Kaleta project in West Africa.[28]

In summary then, it is worth noting that except where a specialised institution exists as in the case of ECOWAS, the other RECs as highlighted in the instance of EAC are mostly involved with the coordination and facilitation of regional energy infrastructure. In the case of energy infrastructure projects, RECs through their Secretariats ensure that various feasibility studies are undertaken and facilitate discussion amongst member states on how to develop various regional energy projects.

Having examined the involvement of RECs in the different types of energy infrastructural projects, in the next section the legal and institutional framework of RECs in relation to the establishment of regional energy infrastructure will be discussed. This is essential in highlighting how RECs operate when involved in the establishment of regional energy infrastructure.

5.5 Role of RECs in the Establishment of Regional Energy Infrastructure

Regional energy infrastructure differs depending on the type of energy being considered. In respect to fossil fuels, the oil and gas pipelines attract less countries as compared to regional electricity interconnections, which

[26] The United Republic of Tanzania and the Republic of Uganda signed a bilateral agreement for the implementation of the Kikagati/Murongo 16 MW hydropower in July 2015. This project is a shared resource on river Kagera between Uganda and Tanzania.

[27] The 10th Meeting of the Sectoral Council on Energy directed the two partner states, that is Uganda and Tanzania, to sign a Memorandum of Understanding and to set up a Joint Technical Committee to spearhead the implementation of the project.

[28] Gnansounou, E., 2008. Boosting the electricity sector in West Africa: An integrative vision. *International Association for Energy Economics, 17*, pp. 23–29.

mostly involve the majority of the neighbouring countries, and in the case of power pooling, almost all the countries in the region are involved.

In this section, through a comparative analysis of EAC and ECOWAS, the role of RECs in regional energy infrastructure establishment and development in SSA as a way of tackling the challenge of energy access is examined. The section specifically looks at the legal and institutional framework of these RECs and it also highlights the legal basis underpinning regional cooperation in the establishment and development of energy infrastructure.

5.5.1 The Enabling Legal Framework for RECs Involvement in Regional Energy Infrastructure

REC involvement in regional energy infrastructure establishment and development should indeed be backed by a legal or regulatory framework for it to be lawful and binding on member states. In this section, through a comparative analysis, the enabling legal framework for EAC and ECOWAS will be examined. On the basis of this legal framework, crucial energy issues pertaining to the protection of energy investments, settlement of disputes, and cooperation in the energy sector among others are covered.

For RECs to be operational, they need to be legally established. In this regard, both EAC and ECOWAS have legal backing having been established under Article 2 of their respective establishing Treaties. Moreover, the establishing Treaties have also emphasised the need for regional cooperation in the development of the energy sector in these regions.[29] It is on this basis that these regions have embraced regional cooperation in the development of energy infrastructure such as power grids and oil and gas pipelines.

Regional cooperation in the development of the energy sector also requires common and harmonised energy policies and programmes: however, full harmonisation of policies has proved hard to achieve in

[29] For the case of ECOWAS, see Article 3(1) & Article26 (1) of the revised ECOWAS Treaty of 1993; for the case of EAC, see Article 5(1) & Article 101 (1) and (2) of The Treaty for the Establishment of the East African Community, 1999.

practice, thus necessitating the simple harmonisation of energy frameworks which can act as a guiding instrument to the various countries in the region. Having this in mind, the establishing Treaties of both EAC and ECOWAS make it mandatory for member states to coordinate and harmonise their energy policies and programmes.[30] On this basis, ECOWAS has established the ECOWAS Regional energy programme and the ECOWAS Protocol on Energy of 2003. The Community energy programme has developed regional coordinated actions at all stages of the energy chain. Comparatively, EAC has not been so active in harmonising energy policies. However, the region attempted to develop an Energy Security Policy Framework and the Cross-Border Electricity Policy, though to date these have not yet been finalised.

The advantage ECOWAS has over EAC regarding harmonisation relates to its common legal framework set up under the auspices of the ECOWAS Energy Protocol. This lays down a common ground on how to handle energy issues in the ECOWAS region. The Protocol took into account the provision of the decision of the Authority of the Heads of States of Governments of ECOWAS relating to the ECOWAS Energy Policy of 1982.[31] Unlike EAC, which currently lacks common rules governing the energy sector, ECOWAS via its Protocol on Energy addresses infrastructural issues concerning transit[32] and it also aims to remove technical, administrative, and other barriers to trade in electricity and other forms of energy in West Africa. It is important to note that the ECOWAS Protocol on Energy has similar provisions with that of the Energy Charter Treaty, which was originally a European initiative. Moreover, both ECOWAS and EAC are observers to the International Energy Charter, a political declaration on energy cooperation.

Notwithstanding the limited and slow development of a regional energy Policy or Protocol at the EAC level, the strength of the establishing Treaty under Article 101 regarding construction of inter-partner state

---

[30] For ECOWAS, see Article 28 (1) of the revised ECOWAS Treaty of 1993.

[31] ECOWAS Energy Policy A/DEC 3/5/82. Decision AIDEC 3/5/82 relating to the adoption of the ECOWAS Energy Policy, adopted by the Authority of Heads of State and Government in 1982.

[32] See Article 7 of the ECOWAS Energy Protocol, 2003.

electrical grid interconnections[33] has witnessed the increasing implementation of decisions concerning the power sector in EAC partner states.[34] These decisions are very high compared to other energy sectors such as fossil fuels and renewables.[35] In a meeting held in March 2016, the Sectorial Council on Energy observed that out of 101 decisions (directives of the Council to the sector), 76% of the power sector decisions were implemented compared to 58% and 55% in fossil fuel and renewable energy sub-sectors, respectively.[36]

Moreover, on the basis of Article 5,[37] the Sectorial Council on Energy approved the Cross-Border Electrification Policy at its 33rd Meeting held on 29 February 2016 and agreed to have the projects identified by the partner states with the involvement of the relevant agencies/institutions. Additionally, at the 10th Meeting, the Sectorial Council on Energy approved the project to develop an Energy Security Policy Framework for the EAC Region.[38] The project is implemented by the EAC in collaboration with the UN Economic Commission for Africa (UNECA) with funding secured by UNECA. The EAC Energy Security Policy is still in draft form, and it was developed to provide guidance for better understanding, measurement, awareness, and management of energy security risks and challenges in the energy sector of the region.[39] Although the development of an Energy Security Policy Framework is a good step in the region, it is not intended to be a binding policy but rather a foundation for EAC as a region or for Partner Sates to devise a legally binding Energy Security Policy in the future.[40]

---

[33] See Article 101 (2) (d) of The Treaty for the Establishment of the East African Community, 1999.

[34] 11th Meeting of the Sectorial Council on Energy Report of the Meeting of the Energy Committee, EAC Headquarters, Arusha, Tanzania, 14–16 March 2016.

[35] Ibid.

[36] Ibid.

[37] Article 5 (1) of The Treaty for the Establishment of the East African Community, 1999, is to the effect that partner states shall develop policies and programmes aimed at widening and deepening cooperation.

[38] See report of the 11th Meeting of the Sectorial Council on Energy, 18 March 2016. REF: EAC/SR/03/2016.

[39] See Draft EAC Energy Security Policy Framework, June, 2016.

[40] Ibid.

Despite having fewer partner states than ECOWAS, the EAC, with only six partner states (in contrast with the 15 in ECOWAS), has been slow in developing regional energy laws. There is a need for EAC to take advantage of its legislative institution to enact regional energy laws and to advocate for the harmonisation of energy laws in the region. Moreover, another advantage that would accrue from such regional energy legislation is the fact that the establishing Treaty of EAC stipulates that regional legislation take precedence over similar national ones.[41] This provision is essential in ensuring that regional decisions are respected and implemented in different partner states, or else it would be fruitless to have regional organisations.

5.5.2 Institutional Framework for RECs in Regional Energy Infrastructure Projects

Typically, RECs are composed of various organs and agencies which together with other institutions are responsible for the performance of the functions of the RECs including enhancing regional energy infrastructural developments.[42] In the case of EAC, these organs are composed of the following: The Summit, the Council, the Co-ordination Committee, and the Sectorial Committee. Similarly, the establishing Treaty of ECOWAS stipulates that the institutions of the Community shall be comprised of the Authority of Heads of State and Government, the Council of Ministers, the Community Parliament, the Economic and Social Council, the Community Court of Justice, and the Executive Secretariat.[43]

In EAC, the Summit is the supreme institution, comprising the heads of states of the partner states.[44] In ECOWAS, these are under the auspices of "The Authority of Heads of State and Government", established under Article 7 of the revised ECOWAS Treaty.[45] Being the highest institution,

[41] Article 8(4) of The Treaty for the Establishment of the East African Community, 1999.

[42] Article 9(4) of The Treaty for the Establishment of the East African Community, 1999.

[43] See Article 6 of the revised ECOWAS Treaty of 1993.

[44] Article 10(1) of The Treaty for the Establishment of the East African Community, 1999.

[45] Article 7 of the revised ECOWAS Treaty of 1993 establishes the Authority of Heads of States as the supreme institution of the community with functions of among others giving direction to the Community.

the Summit/Authority of Heads of States is responsible for the general direction and for providing impetus to the development and achievement of the objectives of the Community, be it in ECOWAS or EAC.[46] The Council is the policy organ comprised of Ministers and it reports directly to the Summit/Authority of Heads of States.[47] The Council is responsible for the functioning and development of the Community and for making policy decisions for the efficient and harmonious functioning of the Community.[48] Moreover, to give effect to the functions of the Council, both EAC and ECOWAS Treaties stipulate that the decisions of Council are binding on the partner states.[49]

Whereas in ECOWAS the establishing Treaty does not explicitly mention other organs such as the Co-ordination Committee and the Sectorial Committee, in EAC these are explicitly provided for in the establishing Treaty. The Co-ordination Committee in EAC is established under Article 17 comprising of the Permanent Secretaries responsible for EAC affairs.[50] Some of the functions of the Co-ordination Committee include submitting reports and recommendations to the Council.[51] Additionally, the Co-ordination Committee also implements the decisions of the Council[52] in addition to receiving reports of the Sectorial Committees.[53] The Sectorial Committees are established by the Council[54] under recommendation of the Co-ordination Committee.[55] These are responsible for

[46] For the case of EAC, the Summit is the one that comprises Heads of States of partner states. See Article 11(1) of The Treaty for the Establishment of the East African Community, 1999; and Article 7 (3) (a) of the revised ECOWAS Treaty of 1993.

[47] In EAC, the Council comprises of Ministers responsible for EAC affairs of each partner state. See Article 13 of The Treaty for the Establishment of the East African Community, 1999; In the case of ECOWAS, the Council comprises of Ministers in charge of ECOWAS Affairs and any other minister of each Member State. See Article 10 (2) of the revised ECOWAS Treaty of 1993.

[48] For EAC see Article 14(3) (a) of The Treaty for the Establishment of the East African Community, 1999; for ECOWAS, see Article 10 (3) of the revised ECOWAS Treaty of 1993.

[49] See Article 16 of The Treaty for the Establishment of the East African Community, 1999; and Article 12 (3) of the revised ECOWAS Treaty of 1993.

[50] Article 17 of The Treaty for the Establishment of the East African Community, 1999.

[51] Ibid, Article 18(a).

[52] Article 18(b) of The Treaty for the Establishment of the East African Community, 1999.

[53] Article 18(c) of The Treaty for the Establishment of the East African Community, 1999.

[54] See Article 14(3) (j) of the Treaty for the Establishment of the East African Community, 1999.

[55] Article 20 of The Treaty for the Establishment of the East African Community, 1999.

the preparation of a comprehensive implementation programme and setting out priorities with respect to the sector in question.[56] Moreover, they are also responsible for monitoring and keeping under constant review the implementation of the programmes of the Community with respect to its sector.[57]

Nevertheless, ECOWAS under its establishing Treaty provides for other organs and institutions with a general competence that ensures they may well have an effect on the functioning of the energy sector. These include the Community Parliament, the Economic and Social Council, the Court of Justice, the Arbitration Tribunal, and the Executive Secretariat.[58] Similarly, just like ECOWAS, the EAC Treaty also established the Parliament,[59] Court of Justice,[60] and the Secretariat[61]; however, the EAC does not mention some organs such as the Economic and Social Council and the Arbitration Tribunal.

With regard to the energy sector, the Sectorial Council on energy comprising energy Ministers of partner states makes decisions affecting regional energy matters including infrastructure; it also directs the Secretariat on what needs to be done in the region, in addition to ensuring the implementation of the Summit directions.[62] With the above established organs and institutions, together with the legal framework discussed in this section, RECs further contribute to the development of regional energy infrastructure by sourcing funding for feasibility studies of the selected regional infrastructure, and these studies are essential to attract investors to take up these regional energy infrastructural projects.

---

[56] Article 21(a) of The Treaty for the Establishment of the East African Community, 1999.

[57] Article 21(b) of The Treaty for the Establishment of the East African Community, 1999.

[58] See Article 13, 14 15, 16 and 17 of the revised ECOWAS Treaty of 1993, establishing the Community of Parliament; Economic and Social Council; the Court of Justice; the Arbitration Tribunal and the Executive Secretariat respectively.

[59] The East African Legislative Assembly is established under Article 38 of the Treaty for the Establishment of the East African Community, 1999.

[60] The East African Court of Justice is established under Article 26 of The Treaty for the Establishment of the East African Community, 1999.

[61] The Secretariat and Staff of the Community are established under Article 47 of The Treaty for the Establishment of the East African Community, 1999.

[62] The Sectorial Council on Energy is advised by the Coordination Committee comprising of Permanent Secretaries of the Ministry of Energy of all the partner states. These are in turn advised by the Energy Committee comprising of technical staff from partner states.

Taking an example from the EAC, in 2013 the Sectorial Council on energy directed the EAC secretariat to put plans in place and source funding to conduct feasibility studies for the oil product pipeline interconnection of Kampala-Kigali-Bujumbura and Mwanza-Isaka-Dar es Salam.[63] In recent years, EAC involvement in sourcing funding for regional energy infrastructure was witnessed in the Mbarara-Mwanza-Isaka-Dar es Salam oil products pipeline project; the EAC Secretariat submitted a proposal to the AFDB for funding of a feasibility study for this pipeline.[64] The project aims at linking Uganda and Tanzania to the existing and future oil products pipeline system as the EAC seeks to link all its partner states with oil products pipelines and to the planned refinery development in Hoima, Uganda.[65]

From the above analysis, the regional efforts taken to establish and develop energy infrastructure in the SSA can be seen. Moreover, this is mostly achieved through the Secretariats, which largely have responsibility for coordinating and facilitating different regional projects. The challenge, however, is that there is lack of capacity in the energy departments of most RECs. In the EAC for instance, a lack of capacity has been evident for several years: in this regard, in their meeting held in 2013, the Sectorial Council raised concerns about the serious capacity constraints as it only had one established staff member and one project staff member. Taking stock of this, the Council directed the EAC secretariat to source funding from development partners to support at least two more senior energy officers. Additionally, the Council requested partner states to consider seconding their staff on a short-term basis to support the Energy department at EAC.[66] In spite of these directions by the Sectorial Council,

---

[63] See report of the 8th Meeting of the Sectorial Council on Energy, 21 June 2013. REF: EAC/SR/165/2013.

[64] As of March 2016, the EAC Secretariat had received notification from the AFDB that a NEPAD-IPPF grant of USD 1,691,130 to finance the feasibility study had been approved. The EAC was expected to jointly raise USD 95,600 as counterpart funding to go towards the facilitation of Steering Committee supervision of the study activities and review of consultant reports. See report of the 11th Meeting of the Sectorial Council on Energy, 18 March 2016. REF: EAC/SR/03/2016.

[65] See report of the 11th Meeting of the Sectorial Council on Energy, 18 March 2016. REF: EAC/SR/03/2016.

[66] See report of the 8th Meeting of the Sectorial Council on Energy, 21 June 2013. REF: EAC/SR/165/2013.

until now the capacity at the EAC energy department is still lacking with only one energy official at the Secretariat.[67] Moreover, in the past it was recognised that the EAC Secretariat faced challenges in convening Sectorial Council meetings as scheduled, which has hindered decision making in the energy sector.[68] Although the Council directed EAC to ensure that it convenes two meetings of the Sectorial Council per year,[69] this has not been delivered on and of course two meetings per year still appears inadequate.

From the discussion above, it is clear that there is potential for RECs to contribute in overcoming the challenge of energy access in SSA through the establishment and development of regional energy infrastructure. However, the general institutional and legal frameworks can be revised and tailored to adequately respond to energy challenges in these regions. Perhaps this could be achieved by setting up specific energy access-related laws and organisations.

5.6 Conclusion

Taking stock of the discussion in this chapter, it would be right to view RECs as being pivotal in ensuring regional cooperation in the establishment and development of regional energy infrastructure including power grid interconnections and oil and gas pipelines. This is true, given their legal and institutional framework which encourages cooperation in the energy sector. Additionally, their role as coordinators and facilitators of regional energy projects also put them in a good position to tackle the challenge of energy access. Taking EAC as an example, it is involved in different coordination facilities including drafting legal documents for regional energy projects and ensuring that the partner states move forward with the project. A case in point is the Nsongezi Hydropower

[67] As of March 2017, the energy department of EAC had only one established staff.

[68] See report of the 8th Meeting of the Sectorial Council on Energy, 21 June 2013. REF: EAC/SR/165/2013.

[69] See report of the 8th Meeting of the Sectorial Council on Energy, 21 June 2013. REF: EAC/SR/165/2013.

Project between the Republic of Uganda and the United Republic of Tanzania: in this project, it is the EAC Secretariat that drafted the Memorandum of Understanding (MOU) for this project, which was sent to the involved partner states for review and signature.[70]

However, in spite of their pivotal role, given the complexity of energy matters and reflecting on the various challenges RECs encounter, including limited capacity, lack of political will, instabilities in some member states, and lack of finances, just to mention a few, one can rightly argue that regional energy matters should be given special attention and specific regional institutions and legal frameworks should be set up to handle these matters, specifically with a focus on energy access. It is true SSA has experienced instances where RECs like ECOWAS have set up specialised institutions to handle energy infrastructural projects, a case in point being WAGP. Perhaps this could be recommended for other RECs like EAC and SADC.

Moreover, just as in other parts of the globe, regional power pools have also been established in SSA as specialised institutions for RECs. Power Pools have become a common mechanism of interchanging power between two or more utilities which provide or generate electricity, with a common agreement to coordinate operations and planning for improved reliability and efficiency. Although power pools are usually concerned with power trading, they play a role with regard to soft regional energy infrastructure development. Additionally, they also facilitate funds for energy infrastructural projects. Taking the example of the West African Power Pool (WAPP), it facilitated the construction of the Nigeria-Benin interconnector Reinforcement Project.[71] There is therefore a need to strengthen regional efforts in establishing and developing energy infrastructure as a way of addressing energy access challenges in SSA.

[70] See report of the 8th Meeting of the Sectorial Council on Energy, 21 June 2013. REF: EAC/SR/165/2013.

[71] African Development Bank: AfDB and WAPP sign US $1.9 million grant for Nigeria-Benin Interconnector Re-enforcement Project, https://www.afdb.org/en/news-and-events/afdb-and-the-west-africa-power-pool-sign-us-1-9-million-grant-for-nigeria-benin-interconnector-reinforcement-project-15689/, last visited on 19 March 2017.

References

Amin, M. (2008, July). Challenges in reliability, security, efficiency, and resilience of energy infrastructure: Toward smart self-healing electric power grid. In *Power and energy society general meeting-conversion and delivery of electrical energy in the 21st century, 2008 IEEE* (pp. 1–5). IEEE.

Armar, A. (2009). *Building regional power pools. A Toolkit* (p. 2). Washington, DC: The World Bank.

Blyden, B. K. (1997). Towards the implementation of an integrated African grid. *IEEE Power Engineering Review, 17*(1), 11–11.

Choi, J. B., Goo, B. K., Kim, J. C., Kim, Y. J., & Kim, W. S. (2003). Development of limit load solutions for corroded gas pipelines. *International Journal of Pressure Vessels and Piping, 80*(2), 121–128.

ECREE: ECOWAS Energy Efficiency Policy (EEEP), September, 2012, P. 8.

Gnansounou, E. (2008). Boosting the electricity sector in West Africa: An integrative vision. *International Association for Energy Economics, 17*, 23–29.

Goldthau, A. (2014). Rethinking the governance of energy infrastructure: Scale, decentralization and polycentrism. *Energy Research & Social Science, 1*, 134–140.

Hector G Sarmiento, Miguel Angel Avila: Issues Associated with International Power Grid Interconnections in Mexico.

Huda, M. S., & McDonald, M. (2016). Regional cooperation on energy in South Asia: Unraveling the political challenges in implementing transnational pipelines and electricity grids. *Energy Policy, 98*, 73–83.

Jerome, A., & Nabena, D. (2016). Infrastructure and regional integration in Africa. In *Region-building in Africa* (pp. 89–108). Palgrave Macmillan US.

Kasumuni, L. (2016, April 30). TZ to spend Sh300bn on buying stake in Uganda refinery. *The Citizen (Tanzania)*. Dar Es Salaam. Retrieved March 5, 2017.

Kennedy, J. L. (1993). *Oil and gas pipeline fundamentals* (p. 2). Pennwell Books.

Nepal, R., & Jamasb, T. (2012). Interconnections and market integration in the Irish Single Electricity Market. *Energy Policy, 51*, 425–434.

Report of the 8th Meeting of the Sectorial Council on Energy, 21 June 2013. REF: EAC/SR/165/2013.

SADC Energy Monitor, 2016. Page 32.

The revised ECOWAS Treaty of 1993.

The Treaty for the Establishment of the East African Community, 1999.

6

Harmonisation of Regional Energy Regulations

6.1 Introduction

The encouragement of regional cooperation in the development of energy resources has been a goal of many inter-governmental organisations not only in SSA but also globally, for example in Europe, Asia, and America. This is not only evidenced in the development of regional energy infrastructure, but also attempts to harmonise energy regulations at the regional level. In Europe this is evidenced by the project of establishing an Internal Energy Market (IEM) for the 28 member states of the European Union.

Regulation of the energy sector and specifically electricity has for many decades been a main objective of many countries globally. This is indeed due to the fact that today's societies rely on electricity for virtually everything. The purpose of electricity has escalated from the original domestic role of lighting to commercial and industrial domains. Typically, power sector reforms, including the establishment of independent regulatory authorities, are undertaken for various reasons including constraining the monopoly power of incumbent suppliers and providing incentives for operating efficiency and quality of services among others. However, it is

V. R. Nalule, *Energy Poverty and Access Challenges in Sub-Saharan Africa*, Energy, Climate and the Environment, https://doi.org/10.1007/978-3-319-95402-8_6

worth noting that regulation is one of the hardest parts of electricity sector reform in many countries, especially due to the fact that it involves monitoring the performance of public and private monopolies.[1]

The subject of this chapter therefore is the regulation of the energy markets in SSA as a means of promoting regional energy cooperation. In recent years, SSA regional organisations have embraced power trading via establishing power pools (e.g. EAPP, WAPP, and SAPP) which essentially enable two or more utilities to exchange power thereby ensuring improved reliability and efficiency. Power pools indeed have a crucial role to play in addressing energy poverty in SSA. However, these were established at the time when regional efforts to regulate the energy sector were not in existence or were still weak. For instance, WAPP has been operational since 2001 but it was not until 2008 that the ECOWAS Regional Electricity Regulatory Authority (ERERA) was established.[2] The same goes for SAPP, which was established in 1995, but it was only in July 2002 that the Regional Energy Regulatory Authority (RERA) came into existence.

In essence, operators control most of the information needed for effective regulation and as such it is essential to establish an institution mandated with harmonising the various regulatory databases and developing methodologies for making comparisons. From a regional perspective, the various regional organisations have a big role to play in establishing regional regulatory bodies. The legal establishment and mandate of such regulatory bodies is key to ensuring the effective harmonisation of various national energy regulations and ensuring regional competition. Given the complexity of regional laws and the difficulty in enforcing them, the issue to be addressed in this chapter is whether harmonisation of regional energy regulations enhances regional cooperation in SSA.

This chapter will consider the above issue in relation to the regulation of the electricity sector in SSA, a key sector in the economies of the

[1] Estache, A., Rossi, M.A. and Ruzzier, C.A., 2004. The case for international coordination of electricity regulation: Evidence from the measurement of efficiency in South America. *Journal of Regulatory Economics, 25*(3), pp. 271–295.

[2] ERERA was established by Supplementary Act A/SA.2/1/08 as a specialised institution of ECOWAS.

region and one that is central to its growth aspirations. To this end a comparative analysis will be employed to discuss issues concerning the following: the legal instruments necessary for achieving the goal of regional regulatory cooperation; the time frame it takes to achieve regional regulatory cooperation; and the main aims underlying the countries' desire to harmonise their energy regulations. The SSA regional organisations, with emphasis on SADC will be compared to more advanced (understood as more complex and longer established) regional organisations such as the EU. Looking at the EU experience, one may ask whether it really offers positive lessons which SSA can draw from in their endeavour to achieve regional regulatory cooperation, or negative lessons.

This chapter employs a three-step framework set out in the following sections. The first section considers the role of National Regulatory Authorities (NRAs) in ensuring successful regional regulatory cooperation. This being a cooperative process, it is essential to determine the commitment of the member states involved, considering that energy is a very sensitive area for many national governments. Section two is concerned with the transition to regional regulatory authorities comparing the experiences in SADC and the EU. It examines the origins, legal instruments, and regulatory gaps to be filled for the successful harmonisation of energy regulations. The chapter concludes by addressing the questions raised above and by stating the regulatory gaps that need to be filled by SSA regional organisations for the attainment of advanced regional regulatory authorities.

6.2 Establishment of National Regulatory Authorities in SSA

Harmonising regional energy regulations is a cooperative exercise that requires support from the member states and NRAs concerned. Historically, it was common practice for governments be it in Europe, America, or Africa to involve themselves in the energy business especially the electricity and gas industries, mainly due to the sensitivity of these industries and their pivotal role in national economic development.[3] However, there has been a major shift in this as evidenced by the introduction of competition and liberalisation of the energy sector in different countries.

To this end, SSA like other regions has experienced a wave of market reforms, as many countries have embraced liberalisation and competition to attract the needed investments in the energy sector. With the liberalisation of the energy sector, the involvement of the private sector has also increased making it necessary for the establishment of autonomous agencies to regulate the sector with the main objective of boosting competition and developing a regional energy policy.[4]

Utility services have a broad range of domestic users who are mostly the voting population, making them political in nature. Regulation is therefore necessary to guard consumers and operators against political opportunism and it also provides incentives for service providers to operate efficiently and make the needed investments.[5] Moreover, most utility services are also characterised by important economies of scale and scope, thus making them very sensitive. Regulation is also essential due to the fact that restructured electricity markets are often characterised by substantial uncertainties, thus deterring new competition in the market, as such regulation is necessary to bring about investor and supplier confidence in the energy sector.[6]

Making a comparison between different countries in the SADC region and the EU, this section aims at analysing whether the legal instruments establishing NRAs enhance regional regulatory cooperation. Prior to addressing this question, this section will discuss issues concerning the legal establishment and objectives of NRAs.

---

[3] Cameron, P.D. and Brothwood, M., 2002. *Competition in energy markets: Law and regulation in the European Union* (p. 5). Oxford University Press.

[4] Bonafé Martínez, E., 2010. *Towards a European energy policy: Resources and constraints in EU law* (Doctoral book).

[5] Pérez-Arriaga, I.J. ed., 2014. *Regulation of the power sector* (p. VIII). Springer Science & Business Media.

[6] Kwoka, J., 2008. Barriers to new competition in electricity generation. *Report to the American Public Power Association, Northeastern University.*

6.2.1 Legal Establishment, Scope, and Duties of NRAs

6.2.1.1 SADC Experience

There are various NRAs that exist in the SADC region and these are outlined in Table 6.1.

These NRAs are established under different legal frameworks and the duties and scope of their powers differ from one country to another. The origin of most of the NRAs in SADC needs to be understood in the context of the liberalisation process in the energy sector, which resulted in major changes to the institutional and legal framework of the energy sector previously characterised by monopolies operating under extensive government control. These changes necessitated the enactment of Energy Regulation Acts, which would later legally establish NRAs.

Taking Zambia's Energy Regulation Board (ERB) as an example, it was established by the enactment of the Energy Regulation Act.[7] Besides the Energy Regulation Act, other governing laws of the ERB include the Electricity Act and the Petroleum Act.[8] The National Energy Regulator of South Africa (NERSA), on the other hand, was established by the enactment of the National Energy Regulators Act, which mandated NERSA to regulate the electricity, piped-gas, and petroleum pipelines industries.[9] It is important to note that there are often other governing laws and in the case of NERSA these include the Electricity Regulation Act, the Gas Act, and the Petroleum Pipelines Act.[10]

Scope of Duties

There are various legal instruments establishing NRAs depending on the country concerned. These legal instruments are also the basis for the

[7] The Energy Regulation Board was established under Section 3(1) of the Energy Regulation Act, Cap 436 of the Laws of Zambia.

[8] The Electricity Act, Chapter 433 and the Petroleum Act, Chapter 434 of the Laws of Zambia.

[9] Section 3 of the National Energy Regulator Act, 2004 (Act No.40 of 2004).

[10] The Electricity Regulation Act, 2006 (Act No. 4 of 2006); Gas Act, 2001 (Act No. 48 of 2001); and Petroleum Pipelines Act, 2003 (Act No. 60 of 2003).

Table 6.1 NRAs and their scope of duties

Country	Regulator	Date of establishment	Scope of duties	Regulatory model
Angola	Regulator Institute for services of Electricity, Water supply and Waste Water Sanitation (HIRSEA)	January 2016[a]	Electricity, water, waste sanitation	Ministry
Botswana	Botswana Energy & Water Regulation (under development)	Under development	Electricity, petroleum, renewable energy	Independent
DRC	Ministry of Mines, Energy & Hydrocarbons/ SNEL	N/A	Electricity	Ministry/utility
Lesotho	Lesotho Electricity and Water Authority (LEWA)	1 May 2013[b]	Electricity, water, and sewerage services	Independent
Malawi	Malawi Energy Regulatory Authority (MERA)	2004[c]	Energy sector[d]	Independent
Mozambique	Energy Regulatory Authority (ARENE)	2017[e]	Electricity	Ministry
Namibia	Electricity Control Board of Namibia (ECB)	2000[f]	Electricity	Ministry
South Africa	National Energy Regulator of South Africa (NERSA)	2005[g]	Electricity, piped-gas, and petroleum pipelines industries	Ministry/ independent

(*continued*)

Table 6.1 (continued)

Country	Regulator	Date of establishment	Scope of duties	Regulatory model
Swaziland	Swaziland Energy Regulatory Authority (SERA)	2007[h]	Electricity	Independent
Tanzania	Energy and Water Utilities Regulatory Authority of Tanzania (EWURA)	2001[i]	Electricity, petroleum, natural gas, and water sectors	Independent
Zambia	Energy Regulation Board of Zambia (ERB)	1997[j]	Energy	Independent
Zimbabwe	Zimbabwe Energy Regulator Authority (ZERA).	2011[k]	Energy	Independent

[a]HIRSEA was recently established in January 2016, replacing the Electricity Sector Regulator Institute (IRSE), which was initially established by Decree No. 4/02 on the Regulatory Institute for Electricity Sector.

[b]The Authority was established through the Lesotho Electricity Authority (LEA) Act No.12 of 2002 as amended. However, from August 2004 to April 2013, it was mandated to only regulate the electricity sector. It was not until 01 May 2013 that it became multi-purpose and also started regulating the water sector.

[c]The Authority was established under Section 3 of the Energy Regulatory Act No. 20 of 2004 as the energy sector wide regulator.

[d]Section 2 of the Energy Regulatory Act defines energy to include both renewable and non-renewable sources of energy.

[e]Law establishing ARENE was recently approved in 2017. ARENE replaces the National Electricity Advisory Council of Mozambique (CNELEC).

[f]ECB was established under the Electricity Act 2 of 2000. This Act has been repealed by the Electricity Act, 4 of 2007.

[g]NERSA was established under the National Energy Regulator Act of 2004. It replaced the National Electricity Regulator (NER).

[h]SERA was established through the Energy Regulatory Act, 2007 (Act No.2 of 2007).

[i]EWURA was established under Section 4 of the Energy and Water Utilities Regulatory Authority of Tanzania Act, CAP 414, Laws of Tanzania.

[j]ERB was established under the Energy Regulation Act of 1995 CAP 436 of the Laws of Zambia, following the issuance of Statutory Instrument number 6 of 1997, the Energy Regulation Act (Commencement Order) of 27 January 1997.

[k]ZERA was established under the Energy Regulatory Authority Act CAP 13:23 of 2011.

scope of duties to be carried out by NRAs. Basically, regulatory authorities can be industry specific, sector specific, or they can be multi-sector with a single regulatory agency for all or most in the infrastructure sector.

On the one hand, some NRAs in the SADC region have limited scope and as such are mandated to only address electricity matters. On the other hand, however, other NRAs in SADC are mandated to regulate a wider range of activities, with some regulating the energy sector in its entirety like the MERA in Malawi.[11] While others in addition to energy are also mandated to regulate the water sector like EWURA in Tanzania.[12]

Some of the advantages associated with multi-sector regulatory authorities are: they are thought to be capable of pooling regulatory resources which could be very advantageous especially in countries with limited regulatory capacity; they are capable of increasing resistance to regulatory capture and political interference and also improving the consistency of decision making across sectors.

It is probable that most NRAs in the SADC region are considering multi-sector regulatory agencies as illustrated in Table 6.1 above. To this end, the Acts establishing NRAs also provide room to widen the scope of activities of NRAs as may be deemed fit by future Acts of Parliament. This is the case with Swaziland and Lesotho, and in this regard the Lesotho Electricity Act of 2002 was amended in 2011 to enable the Regulatory Authority to have regulatory powers over the urban water and sewerage services.[13]

Tasks of NRAs

The legal instruments establishing NRAs in most cases specifically outline the functions and obligations of NRAs. Basically, there are three broad

[11] Malawi Energy Regulatory Authority is established under the Energy Regulatory Act No. 20 of 2004; besides this Act, MERA also operates in accordance to the Electricity Act of 2004; The Rural Electrification Act of 2004 and the Liquid Fuels and Gas (Production and Supply) Act of 2004.

[12] Tanzania's regulatory agency was established under the Energy and Water Utilities Regulatory Authority (EWURA) Act Cap 414 of the laws of Tanzania and it is mandated not only to deal with energy but also the water sector.

[13] The Lesotho Electricity Act No. 12 of 2002 as amended, Laws of Lesotho.

aims of regulatory authorities: to protect consumers from abuse by firms with substantial market power; to support investment by protecting investors from arbitrary governmental action; and to promote economic efficiency. In reality these tasks may not be fulfilled due to various challenges, including political interference, lack of adequate resources, and lack of legal mandate. The functions of NRAs must reflect the scope of legal duties in establishing instruments which are discussed above. Some of the general functions of NRAs that cut across in SADC include the following:

Promotion of the market: with the introduction of liberalisation in energy markets comes the responsibility to ensure effective competition. In Zimbabwe, the regulator is under a legal obligation as stipulated in the Act to promote effective competition within the energy industry.[14] Promotion of competition entails issuing licences to other entities for undertakings in the energy sector as stipulated in the Energy Regulatory Act of Swaziland[15] and in the Regulatory Act of Zimbabwe.[16]

Consumer protection: this takes different forms. In Zambia the regulator is under a legal obligation to design standards with regard to the quality, safety, and reliability of supply of energy and fuels.[17] Consumer protection also entails ensuring that fair prices are charged by licensees, while at the same time making sure that these prices are sufficient to allow licensees to finance their activities and obtain reasonable returns for their efficient operation.[18] With the duty of consumer protection also comes the responsibility to address consumer complaints; in this regard the energy regulator in Swaziland goes further to receive and investigate complaints from consumers on price adjustments made or services provided by any undertaking.[19] It is also in the interest of customers that they have adequate knowledge and education regarding the sector and this is one of the duties of the energy regulator in Malawi.[20]

---

[14] This is a main function for many NRAs such as the Zimbabwe Energy Regulatory Authority under Section 4(g) of the Energy Regulatory Act, 2011 (Chapter 13:23), Laws of Zimbabwe.

[15] Ibid Section 5 (1) (a).

[16] Section 4(e) of the Energy Regulatory Act, 2011 (Chapter 13:23), Laws of Zimbabwe.

[17] Section 6(1) of the Energy Regulation Act, Cap 436 of the Laws of Zambia.

[18] This duty is well stipulated in Section 4(f) of the Energy Regulatory Act, 2011 (Chapter 13:23), Laws of Zimbabwe.

[19] Section 5 (1) (g) of the Energy Regulatory Act, 2007 (Act No. 2 of 2007), Laws of Swaziland.

[20] Energy Regulation Act 2004, Laws of Malawi.

Other duties: there are several other duties carried out by energy regulators such as making recommendations to the relevant Minister as to the measures to be taken through regulations, and this is a core duty of the energy regulator in Zambia.[21] Whereas in some countries such as Namibia, the regulatory legal framework does not make explicit mention of the need to promote cooperation in the SADC region, in Zimbabwe this is addressed as a main function of the regulator. In this case the Act emphasises the need for the regulator to promote co-ordination and integration in the importation, exportation, and pooling of energy from any energy source in the SADC and COMESA regions.[22] Moreover, the regulator is seen as one of the main representatives of the energy industry not only in Zimbabwe but also internationally.[23] The explicit mention of the need for regional cooperation and coordination in national regulatory laws is an encouragement and strong foundation for the harmonisation of energy laws at the regional level.

In carrying out the tasks above, the concepts of independence, enforcement, and accountability are vital in the functioning of NRAs. These concepts are briefly discussed below.

Independence: the independence of regulators is essential in ensuring market confidence, which in return attracts the necessary investment. Regulators should operate independently from political pressures, from ministries, and from regulated enterprises, private or public. To effectively carry out their duties, regulatory authorities must be legally distinct and functionally independent from any other public or private company and must exercise their powers impartially and transparently.[24] These principals are well stipulated in the Acts establishing NRAs in various SADC countries. In South Africa for instance, persons appointed as members of the Energy Regulator must terminate any employment or consultancy with firms or companies engaged in the

[21] Section 6(1) of the Energy Regulation Act, Cap 436 of the Laws of Zambia.

[22] Section 4(d) of the Energy Regulatory Act, 2011 (Chapter 13:23), Laws of Zimbabwe.

[23] Section 4(f) of the Energy Regulatory Act, 2011 (Chapter 13:23), Laws of Zimbabwe.

[24] Bonafé Martínez, E., 2010. *Towards a European energy policy: Resources and constraints in EU law* (Doctoral book) Page 120.

electricity, piped-gas, and petroleum industries.[25] Furthermore, such persons must also disclose to the minister and the Energy Regulator any pecuniary interests they may have in these sectors.[26] This provision in South Africa enhances the concept of independence by ensuring that the regulator's management and staff act independently from any market interest. One concern, however, is whether this independence is not threatened from the interference of the government considering the minister is directly involved in the designation of the members of the NRA. Perhaps it can be argued that this is mitigated by the fact that the Act tries to make the appointment procedure transparent by clearly outlining the qualifications of the members to be appointed by the minister.[27]

Autonomy: This requires the regulatory agencies to have their own resources and funding to reduce the influence of politicians who mostly control the national budgets. The most common method of funding is through levies on the regulated firms or the consumers of regulated services.[28]

Enforcement: The objectives and duties conferred on regulators must be accompanied by enforcement tools at their disposal to actually fulfil their tasks.[29] Regulators like in Zimbabwe are conferred with the duty of arbitration and mediation concerning disputes among and between licensees and consumers,[30] and with these duties comes the necessity to issue binding decisions and being able to enforce those decisions.

Transparency and Accountability: Decisions taken by regulatory authorities should be public provided that confidentiality rules are observed. Transparency also requires clear and simple procedural rules. Some of the

---

[25] Section 6 (3) of the National Energy Regulator Act, 2004 (Act No.40 of 2004), Laws of South Africa.

[26] Section 6 (3) of the National Energy Regulator Act, 2004 (Act No.40 of 2004), Laws of South Africa.

[27] Section 5 of the National Energy Regulator Act, 2004 (Act No.40 of 2004), Laws of South Africa.

[28] Estache, A., 1997. Designing regulatory institutions for infrastructure—Lessons from Argentina. *Public Policy for the Private Sector, 114*.

[29] Bonafé Martínez, E., 2010. *Towards a European energy policy: Resources and constraints in EU law* (Doctoral book) Page 122.

[30] Section 4(f) of the Energy Regulatory Act, 2011 (Chapter 13:23), Laws of Zimbabwe.

most important aspects of transparency and accountability include[31] the following:

- Rules setting deadlines for decisions
- Rules requiring detailed justifications and non-political reviews of decisions
- Processes to ensure that concerned parties have the opportunity to express their views in public hearings and to appeal decisions
- Rules to permit the removal of regulators in cases of proven misconduct

In relation to this, NRAs in SADC such as Energy Regulation Board of Zambia (ERB) and the Institute for Electricity Sector Regulation of Angola (IRSE) allow for judicial review, and this helps to ensure that the decisions are fully reasoned and justified and are also subject to appeals before an independent body other than the government.

Moreover, other issues concerning the design of the agency's decision-making structure should also be considered carefully, including questions relating to the selection of regulators, number of decision makers, funding for the agency, and stakeholder's roles and involvement.[32]

6.2.1.2 EU Experience

In contrast to the SSA approach, where individual efforts of member states played a significant role in the liberalisation of the electricity sector, a regional cooperative approach was employed in the EU to achieve similar goals. This was largely due to the goal of deepening integration that was promoted by EU institutions and agreed to by its member states. As such, the liberalisation of the electricity and gas sectors in the EU was built on a regulatory framework dealing with non-discriminatory network access, free competition on wholesale and retail activities,

[31] Estache, A., 1997. Designing regulatory institutions for infrastructure—Lessons from Argentina. *Public Policy for the Private Sector, 114.*

[32] Smith, W., 1997. Utility regulators: Decision making structures, resources, and start-up strategy.

the protection of consumers, and the fulfilment of public service obligations.[33] The concepts of independent energy regulators and free energy markets are therefore an ultimate outcome of a long process based on the founding Treaty and legislation such as the 1987 Single European Act, which actively promoted integration of national markets into one single European Market. A programme of energy-specific legislation (called Directives and Regulations) resulted from 1996 onwards, in what were called "energy packages" (of which there have been three to date and a fourth under discussion). Establishment of strong energy regulators at national and EU level was deemed to be useful to boost competition,[34] to complete the internal market,[35] and to develop a European energy policy.[36] There was also political support from the European Parliament regarding the establishment of national and regional regulators.[37]

Another thing to note concerns the scope of duties. In contrast to the SSA, where the scope of duties of NRAs stretches across different energy sectors, with some including even the water sector, the NRAs in the EU are generally limited in their scope to the electricity and gas sectors: Chapter IX of Directive 2009/72/EC and Chapter VIII of Directive 2009/73/EC,[38] which assume that regulatory authorities should cover both the electricity and gas sectors.[39] However, the actual design of a national or sub-national regulatory agency is left to the member state

---

[33] Bonafé Martínez, E., 2010. *Towards a European energy policy: Resources and constraints in EU law* (Doctoral dissertation).

[34] Communication from the Commission "Inquiry pursuant to Article 17 of Regulation (EC) No 1/2003 in the European gas and electricity sectors" COM (2002) 841 final.

[35] Communication from the Commission to the Council and the European Parliament "Prospects for the internal gas and electricity market" COM (2006) 841 final.

[36] Communication from the Commission to the European Council and the European Parliament "An energy policy for Europe" COM (2007) 1 final.

[37] P6_TA (2007) 0326 European Parliament resolution of 10 July 2007 on prospects for the internal gas and electricity market.

[38] Directive 2009/72/EC of the European Parliament and of the Council of 13 July 2009 concerning common rules for the internal market in electricity and repealing Directive 2003/54/EC (OJ L 211, 14.8.2009) and Directive 2009/73/EC of the European Parliament and of the Council of 13 July 2009 concerning common rules for the internal market in natural gas and repealing Directive 2003/55/EC (OJ L 211, 14.8.2009).

[39] Recital 33 of Directive 2009/72/EC and Recital 29 of Directive 2009/73/EC.

concerned, which opens up the possibility of diversity similar to that found in the SSA.

6.2.1.3 Analysis

The question to be critically addressed here is whether the legal instruments establishing NRAs have a role to play in ensuring the success of regional regulatory cooperation. We may note that in SSA, many countries rely on national and individual legal instruments to establish NRAs, while in the EU, countries rely mostly on the regional EU legal instruments (which of course are made with considerable input from national governmental bodies).

The argument in this book is that regional cooperation is only possible if there is political will. In this regard, it does not matter whether States rely on national initiatives to introduce regulatory authorities, as is the case in many SSA countries. However, it becomes imperative for these States to respect and commit to move toward regional cooperation in regulating the energy sector. This issue of commitment and political will is indeed central to the success or failure of regional regulatory authorities. In this regard, the inclusion of a provision in the legal instruments establishing NRAs regarding the willingness and commitment to work with and contribute to regional regulatory authorities is essential. This provision in NRA establishing legislation will ensure that NRAs take harmonisation of energy laws at a regional level as a legal obligation and as such it will influence political support on the ongoing harmonisation efforts in various SSA regional organisations.

Another important thing to note with the EU experience is the time frame taken for the EU to achieve the Internal Energy Market (IEM) project. This raises a question as to whether the number of countries involved plays a significant role in the achievement of effective regional regulatory cooperation. We note that the IEM project was launched as an idea as much as three decades ago in 1988. During the subsequent period, the number of member states has grown from 12 to 28 (and soon declining to 27). Yet the most recent set of proposals (the so-called Clean Energy Package) underlines the fact that a complete IEM has not yet been achieved. Would the experience be different for regional organisations such as the East African Community (EAC) which have only 6 member states or SADC and ECOWAS with their 15 member states each?

The number of countries involved may of course pose some challenges, but we note that in the SSA, the movement towards liberalisation and subsequently the establishment of regulatory authorities is still in its infant stages for many countries. As indicated in the table above, not many NRAs in SADC are even a decade old and most of them are still undergoing reforms. This has a significant impact on the success of regional cooperative mechanisms in energy regulation. Some countries such as Botswana have not even established regulatory authorities, and other countries such as Angola and Lesotho have reformed their NRA to make them multi-purpose.[40] Definitely, if there were only a few NRAs involved (we might posit a number such as five), then the process of harmonisation of energy laws could be expected to be easy. However, since power pooling mostly involves numerous countries, my recommendation in this regard is for regional organisations in SSA to put strict measures in place, which ensures that the various countries are accountable with regard to their progress or lack of it in harmonising regional energy laws. For instance, countries which have not yet established NRAs should be encouraged by peer pressure to establish such bodies.

Regarding the scope of duties, the question to ponder on is whether the scope of duties of NRAs affects their ability to cooperate at a regional level. Could it be that NRAs which are only focused on electricity in SSA can easily fulfil regional requirements compared to NRAs which have a wide scope of duties ranging from energy to water as is the case in Tanzania? Could the achievements of the EU be attributed to the uniform scope of duties of NRAs in the member states, and if so should SSA regional organisations follow the same example?

In responding to the issues above, it is imperative to note that regional bodies cannot control or influence the scope of duties of NRAs; however,

---

[40] Angola's Regulator Institute for services of Electricity, Water supply and Waste Water Sanitation (HIRSEA) and Lesotho's Electricity and Water Authority (LEWA) were recently made multipurpose and as such they regulate both electricity and water.

at a regional level the focus should be on specific sectors such as electricity and renewable energy. Other sectors like water and sanitation, which are mostly included in the various NRAs in SADC, should be dealt with separately at a regional level, that is they should not be under the energy sector. Fortunately, many SSA regional organisations such as EAC and SADC already have separate sectors for water which are different from the energy sector. However, as stated in Table 6.1, numerous NRAs are expanding their scope of duties to include not only the energy sector but also water and sanitation.

6.3 Regional Harmonisation of Energy Regulations

The establishment of NRAs in member states is not a guarantee that they will cooperate at a regional level, and as such the establishment of regional regulatory authorities seems inevitable to harmonise energy regulations by bridging the gap and differences which may exist amongst the various NRAs. Although this chapter is mostly focused on the power sector, the same sector is undergoing substantial reform including the transition towards smart and clean energy, which definitely necessitates changes in the traditional design of the energy sector including the regulatory frameworks.[41]

Some of the new developments in the electricity sector which definitely have an impact on regulation include among others the reality that energy consumers are now able to produce energy themselves, for instance by installing solar panels and supplying it to the energy network and thus becoming prosumers.[42] This indeed implies that regulation in the energy sector both at national and regional levels should be able to address these new developments. Moreover, developing countries which for the past decades and centuries were mostly reliant on traditional energy such as biomass have now experienced improvements, and there

[41] Lavrijssen, S. and Carrillo, A.J., 2017. Radical innovation in the energy sector and the impact on regulation.

[42] Ibid.

are now visible efforts and progress to shift to clean and modern energy such as electricity and renewables. In this regard, there is a need to reflect on these changes concerning the various transitions, reforms, and progress in the national and regional energy regulations of SSA countries and regional organisations.

In this section, SSA regional regulatory authorities specifically in SADC will be discussed. A comparative analysis is employed comparing the EU and SSA experiences regarding the establishment and functioning of regional regulatory authorities.

6.3.1 Establishment of Regional Regulatory Authorities

The need for regional regulatory authorities is inevitable in regions which have undergone the establishment of power pools. Power pools are generally essential for the successful operation of a competitive electricity market. Besides minimising generation costs and improving conditions on the supply side, other advantages of power pools include the following:

- They provide a source of firm back-up and top up to support either a generator or supplier offering long-term contracts to final customers.
- They provide a ready market for generators unable to sell their power under a contract or wanting a market for spill or excess production.
- They provide a reference price to be used in signalling the optimal development of generation and transmission capacity on the system.[43]

It is due to the enormous advantages associated with power pools that they have been established in different parts of the globe. These power pools are set up with the mission of promoting equitable sharing of responsibilities for planning, developing, and monitoring the technical hardware and software infrastructure required for the safe, reliable, and cost-effective integration of national power grids.

[43] Hogan, W.W., 1995. *Coordination for competition in an electricity market*. Cambridge, MA: John F. Kennedy School of Government, Harvard University.

Typically, prior to the establishment of power pools and other regional coordination activities, there are some organisational activities that must be undertaken. These include among others, the preparation, negotiation, and adaptation of the Intergovernmental Memorandum of Understanding and the Inter-utility Memorandum of Understanding. The Intergovernmental Memorandum of Understanding basically grants permission for the utilities to make a contract and it also provides guarantees regarding obligations resulting from an interconnection contract. The Inter-utility Memorandum of Understanding is among national utilities, and it defines ownership of assets and other key principles to be followed on establishing, putting in place, and enforcing technical planning, operators, and commercial aspects of regional power system integration.

Besides setting up the legal and regulatory framework, there are also other key factors which are crucial such as the ability of the national utilities to maintain flexibility in the setting up of structures which can enable them to plan for and implement cross-border interconnection facilities and also to harmonise the operational rules of practice for their interconnected national power grids.[44]

With power pool coordination comes the need to harmonise the various energy regulations of the countries concerned. Defining a common legal or regulatory framework to achieve regional objectives is key to ensuring the success of power pools.

Harmonisation of energy regulations at the regional level is hard to achieve and this is due to the fact that these sectors are not only sensitive but also highly influenced by national politics. Moreover, the close association of most energy utilities with their respective governments makes it hard to not only achieve full liberalisation but also to achieve harmonisation of the regulations at the regional level. Additionally, not many NRAs are willing to cede their powers to regional regulatory authorities and as such it has become hard to realise full harmonisation. This has in turn forced some regional regulatory authorities in opting to only set guidelines at the regional level while leaving the entire decision making

[44] World Bank, 2008. *Building regional power pools: A toolkit. Technical report*. Washington, DC: World Bank.

in the hands of NRAs. This section will focus on the RERA, a regulatory body in the SADC region, which will be compared with the EU Agency for the Cooperation of Energy Regulators (ACER) in as far as the comparison addresses issues relevant to this chapter.

6.3.1.1 Brief Background of SADCs Regional Electricity Regulatory Authority

The RERA is a formal association of independent electricity regulators in Southern Africa whose membership is open to the electricity supply industry regulatory bodies in each country within SADC. Prior to discussing the brief background of RERA, it is essential to note that RERA was established to ensure the effective operation of the SAPP, which is discussed briefly below.

Southern African Power Pool (SAPP)

SAPP is a regional energy market in Southern Africa established by the signing of the Intergovernmental Memorandum of Understanding in 1995. Other relevant documents which govern SAPP include the Inter-Utility Memorandum of Understanding, which established SAPP's basic management and operating principles; Agreement between Operating Members which established the specific rules of operation and pricing; and the Operating Guidelines, which provide standards and operating guidelines.[45] Supplying reliable and economical electricity to customers of member states is the main objective of SAPP. It is also worth noting that before SAPP came into existence, SADC countries had already expressed interest to cooperate in the electricity sector as far back as the 1950s when a line between Nseke in the DRC and Kitwe in Zambia was constructed in 1958 to supply electricity to the Zambian copper mine. To ensure that SAPP effectively fulfilled its objective of coordinating the planning and operation of the electric power system among member

[45] See SAPP Annual Report, 2015, http://www.sapp.co.zw/annual-reports, last visited on 3 October 2017.

states, the Coordination Centre Board was created in 2007 to govern the activities of the SAPP Coordination Centre.

Even though SAPP in itself passes as a forum to provide for regional solutions to electric energy problems, the need for a regional regulatory body to harmonise energy regulations among the utilities which are members of SAPP is crucial for the success of SAPP; this is the reason underlying the establishment of RERA.

Regional Regulatory Authority (RERA)

The legal basis for the establishment of RERA is the terms of the SADC Protocol on Energy (1996), the SADC Energy Cooperation Policy and Strategy (1996), the SADC Energy Sector Action Plan (1997), and the SADC Energy Activity Plan (2000). These laws expressly state the need for cooperation in the energy sector and lay the foundation for regional regulatory cooperation.

In addition to the regulatory framework, the origin of RERA also needs to be understood in the context of the desire for regional cooperation expressed by national regulatory authorities in Southern Africa. Although the official establishment of RERA by the SADC Ministers responsible for energy occurred in 2002, the commitment for regional cooperation in the regulatory field dates back to August 1998 when the Energy Regulation Board of Zambia (ERB) and the then National Electricity Regulator (NER) of South Africa organised a workshop in Johannesburg, South Africa on regional regulatory cooperation in SADC countries. This was followed by another workshop held in Gaborone, Botswana in March 2000, wherein a process to create a regional consultative platform for electricity regulatory matters in SADC was agreed on and this was later followed by the official establishment in 2002.

Over recent years there have been developments in the state of regulation in SADC and it is evident that different countries are making efforts to embrace reforms in the energy sector in general and utilities in particular. At the launch of RERA, there were only four founding members. However, since then, the Association has witnessed an increase in national regulatory authorities and to date 11 out of the 15 SADC Members States are represented at RERA.[46]

This increase in RERA membership clearly illustrates the commitment for regional regulatory cooperation in SADC but the question arises as to how to make this commitment a reality. Answering this requires examining the experiences of other regulatory associations outside of Africa in order to analyse the regulatory gaps that need to be filled to make regional regulatory cooperation and the harmonisation of energy laws a reality in Southern Africa and other African regional organisations.

6.3.1.2 A Comparative Analysis of the Objectives and Jurisdiction of Regional Regulatory Authorities

Typically, Regional Regulatory Authorities are created with the main purpose of harmonising energy regulations among the member states or utilities involved. This may take the form of a voluntary association whose decisions are not binding or an independent authority whose decisions are binding on member states. In the case of RERA, it is effectively a consultative platform with three main strategic objectives: facilitating capacity building and information sharing among member states at both national and regional levels; facilitation of electricity supply industry policy, legislation and regulations; and ensuring regional regulatory cooperation. The principle of subsidiarity which advocates for a degree of independence among member states applies in RERA and this is due to the fact that the NRAs are left with all the powers concerning regulation and the regional regulator only comes in for consultation and recommendations.

---

[46] These include: Swaziland Energy Regulatory Authority (SERA); Electricity Control Board of Namibia (ECB); Energy and Water Utilities Regulatory Authority of Tanzania (EWURA); Energy Regulation Board of Zambia (ERB); Institute for Electricity Sector Regulation of Angola (IRSE); Lesotho Electricity and Water Authority (LEWA); Malawi Energy Regulatory Authority (MERA); National Electricity Advisory Council of Mozambique (CNELEC); National Energy Regulator of South Africa (NERSA); Zimbabwe Energy Regulator Authority (ZERA). The remaining four countries are at various sector reform states that include plans to establish independent energy regulatory agencies. These countries include Botswana, DRC, Mauritius, and the Seychelles.

In an endeavour to achieve the objectives stated above, RERA regulatory guidelines for cross-border power trading in Southern Africa were established. The overall objective of the guidelines is to ensure that efficient cross-border deals are not constrained by unclear or complicated processes for making regulatory decisions.

These guidelines include:

1. Regulator's powers and duties in cross-border trading
2. Working to ensure compatible regulatory decisions
3. Timing of regulatory interactions for proposed cross-border transactions
4. Licencing cross-border trading facilities imports and exports
5. Approving cross-border agreements in importing countries
6. Approving cross-border agreements in exporting countries
7. Approving cross-border agreements in transit countries
8. Approving transmission access and pricing and ancillary services
9. Promoting transparency in the regulation of cross-border trading

Besides organising trainings for the staff of the different NRAs in SAPP, the establishment of these guidelines can also be considered as one of the achievements of RERA so far. However, these guidelines alone cannot effectively ensure the achievement of the objective regarding regional regulatory cooperation and the harmonisation of energy laws.

Looking at other regional regulators, it is worth noting that there are many aspects that are crucial for the success of these regional bodies. The Federal Energy Regulatory Commission (FERC), for example, is an independent regulatory agency within the US Department of State. FERC regulates the interstate transmission of electricity, natural gas, and oil, and its independence and ability to self-fund its projects have indeed contributed to its success. In contrast, RERA is only a formal association of independent regulators and this implies that it does not have any form of independence and legal mandate. As such it can only act as an advisory body.

Experience from other regional organisations such as the EU has shown that regulatory gaps can delay the achievement of regional regulatory cooperation. Looking briefly at the EU experience, we note that just like SADC, it also started with a voluntary association under the auspices of the Council of European Energy Regulators (CEER), which existed before the establishment of ACER. ACER is briefly discussed below:

The Agency for the Cooperation of Energy Regulators (ACER)

ACER is an EU Agency created to further progress the completion of the internal energy market both for electricity and natural gas, and to foster cooperation among European energy regulators. Cooperation between national regulatory authorities was considered essential in building a European energy market. There is indeed a legal duty for regulatory authorities to cooperate at a regional level regarding cross-border issues[47] and also to cooperate in overseeing the operation of European grids and facilitate competition within the internal market.[48]

The establishment of ACER can further be understood by the need to cover a regulatory gap in the first energy package, which neglected to cover the development of interconnectors and cross-border regulatory issues, despite the fact that a transition to a fully operational IEM required close cooperation between member states, national regulators, industry, community institutions, consumers, traders, and other interested parties.[49] Therefore, in an effort to cover this regulatory gap, a group of national regulatory authorities in 1997 started to meet informally together with the Commission and the industry in Florence, for the Electricity Regulatory Forum (ERF), and in Madrid for the Gas Regulatory Forum (GRF). The ERF was initially convened by the European Commission to monitor and discuss implementation of the Electricity Directive. The principal issues addressed in the meetings included cross-border tarification, allocation of capacity, and congestion management. The Forum was composed of participants from member states, The European Commission, ETSO, representatives from the

---

[47] Article 37 (1)(c) of Directive 2009/72/EC.

[48] Bonafé Martínez, E., 2010. *Towards a European energy policy: Resources and constraints in EU law* (Doctoral book) Page 123.

[49] Cameron, P.D. and Brothwood, M., 2002. *Competition in energy markets: Law and regulation in the European Union* (p. 284). Oxford University Press.

European Parliament, the network users, traders, consumers, and pools.[50] The GRF on the other hand was established to assist in the detailed implementation of the Gas Directive. The main objective of GRF was to encourage a greater convergence of view, regulatory practices, and actions among participants as the liberalisation process gained momentum.[51]

A major concern behind the establishment of the fora was the possibility that liberalisation within the EU might lead to the creation of 15 liberalised, but largely separate, national markets in electricity and gas instead of an integrated EU energy market.[52] Although the two forums achieved some successes, their voluntary and informal character made them vulnerable in the face of intransigence on the part of incumbent market players.[53]

Following the forums, CEER was established in 2000 to ensure further cooperation between national regulators. CEER integrated the European Commission under a more formalised structure called the European Regulatory Group for the Electricity and Gas (ERGEG) in the context of the second energy package.[54] However, agreements were mainly reached on a voluntary basis, and this pragmatic way of cooperation between regulatory authorities slowed down at the moment when the liberalisation deadlines were nearing.[55] As a result, the need to create a new European energy regulator was recognised by the Commission and was supported by national regulators. The Commission presented the proposal for the establishment of ACER according to the operational framework for European agencies developed in the draft inter-institutional agreement.[56]

---

[50] Cameron, P.D. and Brothwood, M., 2002. *Competition in energy markets: Law and regulation in the European Union* (p. 286). Oxford University Press.

[51] Ibid. Page 301.

[52] Cameron, P.D. and Brothwood, M., 2002. *Competition in energy markets: Law and regulation in the European Union* (p. 284). Oxford University Press.

[53] Cameron, P.D. and Brothwood, M., 2002. *Competition in energy markets: Law and regulation in the European Union* (p. 301). Oxford University Press.

[54] Bonafé Martínez, E., 2010. *Towards a European energy policy: Resources and constraints in EU law* (Doctoral book) Page 124.

[55] Bonafé Martínez, E., 2010. *Towards a European energy policy: Resources and constraints in EU law* (Doctoral book) Page 124.

[56] Cameron, P.D., 2005. Legal aspects of EU energy regulation: Implementing the new directives on electricity and gas across Europe. Bonafé Martínez, E., 2010. *Towards a European energy policy:*

Lessons for RERA

The experience of the EU regarding the establishment of ACER proves that SADC has a long way to go to ensure that RERA can effectively achieve its objectives. Issues regarding the extent and powers of RERA, jurisdiction, scope of duties, and legal mandate are crucial for its success. Currently, RERA has no rule making authority and as such can only provide guidelines and recommendations as it has no capacity to issue binding decisions. It is also dependent on SADC and the SAPP Coordination Centre for the approval of decisions and solutions. Comparatively, ACER's authority stretches from issuing non-binding opinions and recommendations to issuing binding individual decisions especially in cases related to cross-border infrastructure issues. Moreover, the organisation and staffing of ACER also play a big role in ensuring that it effectively fulfils its objectives. ACER is composed of both permanent staff and experts seconded from the various NRAs and it has a Board of Regulators who oversee its regulatory activities.[57]

6.4 Conclusion

There are many benefits that can be enjoyed from regional regulatory authorities such as increasing the integration of infrastructure systems, harmonisation of market structures and the sharing of expertise, experiences, and information. Achieving these benefits will not only require a lot of commitment from the member states involved, but also strong legislation and a legal mandate of the regional regulatory bodies to ensure compliance on the part of the parties involved.

A lack of legal status and authority undermines the progress or success of any institution. Currently, under RERA's establishing legal framework,

---

Resources and constraints in EU law (Doctoral book): Eberlein, B., 2005. Regulation by cooperation: The "third way" in making rules for the internal energy market. *Legal Aspects of EU Energy Regulation*, pp. 59–88.

[57] These Boards of Regulators are composed of senior representatives of NRAs. See Agency for the Cooperation of Energy Regulators: Organisation, http://www.acer.europa.eu/en/The_agency/Organisation/Pages/default.aspx, last accessed on 3 October 2017.

it is only an advisory association whose decisions are not legally binding to NRAs. Additionally, most NRAs in SADC are still in their infant stage with little experience and expertise in the regulation sector. Moreover, some member states have not yet set up NRAs. This necessitates the need for a strong regional regulator with experienced experts and staff to guide the regional efforts of power trading as represented by SAPP. In this regard, there is a need to revise the legal mandate of RERA to make it an Authority that can issue legally binding decisions to NRAs.

Furthermore, given the fact that most NRAs in the SADC region are now considering widening their scope of jurisdiction to other energy sectors, there is a need for the regional body to also widen its scope to include petroleum activities including oil and gas. Additionally, the legal mandate should include the regulation of renewable energy at the regional level. There is no doubt that renewables now have a big role to play in tackling the challenge of energy access in SSA. In this regard, most member states have legislated on renewables and there are regional efforts to establish regional bodies mandated with promoting renewable energy in SSA. Moreover, technological and economic developments such as solar panels have changed the traditional way of viewing the electricity sector, as these now empower energy consumers to also contribute to the production of energy through the panels; and costs are falling. There is therefore a need for a legal mandate to widen the scope of jurisdiction of RERA to also include the regulation of renewables, and this should take into consideration the various private companies in SADC which are now actively investing in the renewable energy sector.

Besides expanding the legal mandate of RERA, the establishing law should also emphasise and provide for the principles of independence, autonomy, transparency, and accountability of RERA. These principles should apply to RERA in the same way that they apply to the NRAs. Staffing at RERA should also be improved by ensuring that seconded experts from concerned NRAs are invited in addition to the permanent staff. Moreover, a Board of Regulators (similar to the one in ACER) should also be established in RERA to oversee its activities. Another organ which will have to be established once RERA becomes an Authority is the Board of Appeal, which should be independent from RERA. This is essential to ensure that those not content with RERA decisions have an opportunity to appeal such decisions.

Another recommendation is to make RERA self-funding (similar to FERC), so that it is entitled to raise revenue through annual charges to the natural gas, oil, and electric industries it regulates. Financial independence will reduce the likelihood of interference from regional and national politicians.

References

Bonafé Martínez, E. (2010). *Towards a European energy policy: Resources and constraints in EU law* (Doctoral book) page 124.

Cameron, P. D. (2005). Legal aspects of EU energy regulation: Implementing the new directives on electricity and gas across Europe.

Cameron, P. D., & Brothwood, M. (2002). *Competition in energy markets: Law and regulation in the European Union*. Oxford University Press.

Communication from the Commission to the European Council and the European Parliament "An energy policy for Europe" COM (2007) 1 final.

Directive 2009/72/EC of the European Parliament and of the Council of 13 July 2009 concerning common rules for the internal market in electricity and repealing Directive 2003/54/EC (OJ L 211, 14.8.2009).

Eberlein, B. (2005). Regulation by cooperation: The "third way" in making rules for the internal energy market. *Legal aspects of EU energy regulation*.

Electricity Regulation Act, 2006 (Act No. 4 of 2006); Gas Act, 2001 (Act No. 48 of 2001) and Petroleum Pipelines Act, 2003 (Act No. 60 of 2003).

Energy and Water Utilities Regulatory Authority of Tanzania Act, CAP 414, Laws of Tanzania.

Energy Regulatory Act, 2007 (Act No. 2 of 2007), Laws of Swaziland.

Energy Regulation Act 2004, Laws of Malawi.

Energy Regulation Act, Cap 436 of the Laws of Zambia.

Energy Regulatory Act, 2011 (Chapter 13:23), Laws of Zimbabwe.

Estache, A. (1997). Designing regulatory institutions for infrastructure—Lessons from Argentina. *Public Policy for the Private Sector, 114*.

Estache, A., Rossi, M. A., & Ruzzier, C. A. (2004). The case for international coordination of electricity regulation: Evidence from the measurement of efficiency in South America. *Journal of Regulatory Economics, 25*(3), 271–295.

Hogan, W. W. (1995). *Coordination for competition in an electricity market.* Cambridge, MA: John F. Kennedy School of Government, Harvard University.

Kwoka, J. (2008). Barriers to new competition in electricity generation. *Report to the American Public Power Association, Northeastern University.*

Lavrijssen, S., & Carrillo, A.J. (2017). Radical innovation in the energy sector and the impact on regulation.

Lesotho Electricity Act No. 12 of 2002 as amended, Laws of Lesotho.

National Energy Regulator Act, 2004 (Act No.40 of 2004), Laws of South Africa.

Pérez-Arriaga, I. J. (Ed.). (2014). *Regulation of the power sector* (p. VIII). Springer Science & Business Media.

Recital 33 of Directive 2009/72/EC and Recital 29 of Directive 2009/73/EC.

SAPP Annual Report, 2015.

Smith, W. (1997). Utility regulators: Decision making structures, resources, and start-up strategy.

World Bank. (2008). *Building regional power pools: A toolkit. Technical report.* Washington, DC: World Bank.

7

Conclusion

7.1 Conclusions

There is a clear trend in the economic, political, and social spheres to create supranational or regional and international entities and institutions generally. The aspects of regionalism and globalism were marked features in the past and till now they have played a key role in different sectors, including energy, although with the caveat that there have been some recent threats to these notions as elaborated in the preceding chapters.

In the energy sector, regionalism and globalism have been experienced not only in the developing world but also in the developed world. Supranational or regional energy entities such as power pools, regional regulatory authorities, regional energy infrastructure institutes, and regional and international energy organisations have clearly echoed the success of the notions of regionalism and globalism.

As explored throughout this book, regional cooperation in addressing the challenge of energy access is associated with various advantages such as reducing the cost of providing electricity (in the case of power pools), reducing the cost of establishing energy infrastructure (in the

V. R. Nalule, *Energy Poverty and Access Challenges in Sub-Saharan Africa*, Energy, Climate and the Environment, https://doi.org/10.1007/978-3-319-95402-8_7

case of regional energy infrastructure), providing a standardised legal framework (in the case of regulatory authorities), and reducing the risk of climate change (in the case of regional development of renewable energy). However, realising the benefits of regional cooperation in the energy sector requires a thorough understanding of the different cooperating mechanisms and institutions in addition to a robust legal framework.

Whereas the benefits of regional cooperation in the energy sector are more visible in the developed world, partly due to the advanced stages of such regional projects, on the contrary, little progress can be seen in the developing world due to the immature stage of such projects coupled with barriers such as financial constraints, weak legal and institutional frameworks, and political instabilities to name but a few. As a result, the full potential of the benefits that accrue from regional cooperation cannot be fully realised due to the various obstacles.

In this book, the previous chapters have demonstrated how regional cooperation can be applied in SSA to overcome the challenge of energy access. A detailed study of the various regional cooperative mechanisms in the energy sector was undertaken in order to point out their potential success and challenges in SSA. Various questions have been raised in the previous chapters and some of these received responses, albeit tentative at times. This concluding chapter will now set out the final findings of this book together with contributions and recommendations.

7.2 Findings

In undertaking this research, the study was motivated and particularly concerned with the escalating problem of energy access in SSA despite the numerous initiatives over the years to address the problem. Taking into cognisance the fact that most SSA countries are faced with common energy challenges, this inspired me to analyse the regional cooperative forums in place to tackle the challenge of energy access and how these are functioning. The findings from this effort are presented in the next section.

7.2.1 Lack of Coordination on International Initiatives

As discussed in Chap. 1, there are several initiatives to tackle the challenge of energy access in SSA but unfortunately these are not well coordinated and as such there is a lot of duplication in these initiatives. At an international level, there exist several such initiatives from international institutions including the World Bank, UN, USAID, DFID, and the G20 just to mention but a few. Unfortunately, there is no common mechanism to coordinate these various initiatives and there is no way to see if these have been effective in solving the problem. Furthermore, at the regional level, regional organisations have set up energy cooperative institutions, but again these have proved to be not well coordinated.

The lack of coordination can also be illustrated by the failure to come up with an agreed definition of energy access, implying that different international institutions view the problem differently and as such apply different solutions as they deem fit. As already emphasised in the previous chapters, misconceived problems often lead to misconceived solutions. Therefore, this lack of coordination has indeed contributed to the escalation of the problem of energy poverty in SSA.

7.2.2 Weak Regional Energy Legal and Institutional Frameworks

The desire to cooperate in the development of the energy sector and tackle the challenge of energy access already exists and is already visible in the various SSA regional organisations. The main challenge is with regard to legal and institutional frameworks to drive this desire and make it a reality. As illustrated in Chaps. 4, 5, and 6, regional institutions such as regional regulatory authorities, power pools, and renewable centres are not strong enough to achieve the objectives of the regional organisations with regard to cooperation in the energy sector. Moreover, lack of finances and staffing has also made these efforts fruitless. Additionally, regional efforts are also constrained by the fact that some member states are still struggling to set up the necessary legal and institutional frameworks in their countries.

7.3 Recommendations

From a study of regional cooperation in addressing the challenge of energy access in SSA, the research has identified four common issues and challenges that urgently require thorough attention by the relevant countries, regional organisations, and international organisations in order to ensure that the continent is fully electrified. These issues/challenges include the following:

1. Clearly defining and understanding key concepts

   The main concept to be understood and agreed upon is that of "energy access". As illustrated in Chap. 1, there is currently no clear consensus on what energy access means. Different international institutions, development partners, and scholars have defined energy access differently.

   The contribution of this book with regard to the definition of energy access is elaborated in Chap. 2 of this book and in summary it is as follows: The definition of energy access should be in some phases, the first phase focusing on "household having access to electricity". Electricity for businesses and institutions should be under a different phase. In this definition, the focus is on the element of accessibility only and other elements such as affordability, reliability, and sustainability should be tackled separately for reasons already set out (see Chap. 2).

   The reasoning behind this definition is that the same measures cannot be employed to tackle such different aspects. The issue concerning affordability, for instance, is very personal and has a lot to do with the economic situation in the country, which is sometimes influenced by the political situation of that country. These can hardly be controlled by regional or international energy players. Moreover, measures aimed at eradicating poverty should be geared towards addressing the issue of affordability because it is possible for people to have access to reliable and affordable electricity without really being able to afford it. This is true considering that a number of people live well below the poverty line in SSA, depending on only one dollar per day. There is a need to take such people into consideration when discussing energy access, and this can only be done by separating out the key issues so that each issue or aspect gets a specific solution.

The issue of accessibility, in contrast, has a lot to do with the investments in the country. For example, is there sufficient energy infrastructure such as electricity transmission lines to enable access to electricity? This brings into play the need for more hard infrastructure in these countries, and in this regard the various international institutions which are interested in solving the problem of energy access should invest in such hard infrastructure.

Once there is hard infrastructure in place, and the majority of people in rural and urban areas can access the electricity, then the issue of reliability can be tackled. It is imperative to note that international, regional, and national efforts should separate these elements because they are different and require different solutions. Mixing them up means that the initiatives taken to solve the challenge will not be effective. In this regard, this research suggests a more focused definition of energy access which can lead to focused solutions, which can be attained through regional cooperation.

2. Strengthening the regional legal and institutional framework

   As illustrated in the book, some SSA regional organisations do not have a specific law in place regarding energy cooperation. The suggestions and contributions in this regard are as follows:

- Regional Regulatory Authorities

  In Chap. 6, the use of RERA as a case study illustrated the challenges facing SSA regional regulatory authorities. I make recommendations regarding the need to make RERA a legal and functional entity with autonomous powers to regulate the SADC energy sector.

  EAC Energy Act

  EAC should use its legislative organ to enact an Energy Access Bill (later to become an Act), which can take into consideration all the provisions of an Energy Protocol. This Bill should emphasise the need for regional energy cooperation including making provisions for the autonomy and independence of regional energy institutions in the EAC. Once enacted into law, the EAC Energy Act should be adopted by the member states and it should also take into consideration new technological developments in renewable energy.

- Establishment of an African Energy Access Organisation

As discussed in the previous chapters, there is a lack of coordination with regard to tackling the challenge of energy access in SSA. In this regard, there is a suggestion for SSA countries to come up with a specific African organisation dedicated to tackling the challenge of energy access, just as the Europeans and Asians did when faced with the challenge of energy security, which necessitated the establishment of IEA in Europe and OPEC in Asia.

Africa needs the same specific organisation perhaps by the name of "African Energy Access Organisation (AEAO)". This organisation should attract all the energy poor countries of SSA. Amongst its legal mandate should be the coordination of the various efforts and initiatives of international institutions, development partners, and donors to avoid the problem of duplication of capacity and aims, which is currently rampant throughout the African continent. AEAO should also place emphasis on practical solutions such as hard energy infrastructure (including electricity transmission lines, oil and gas pipelines, and oil refineries). It should also be able to coordinate the various regional efforts in the energy sector as witnessed by various forums such as power pools and regulatory authorities. The composition should consist of representatives from the various SSA regional organisations.

- Establishment of specialised regional energy infrastructure institutions

Specialised institutions in both electricity grids and oil and gas infrastructure should be established in the various SSA regional organisations. As highlighted in Chap. 4, if the role of regional organisations in addressing the challenge of energy access is to be more effective, then there is a need to enhance regionalism in the development of oil and gas infrastructure in SSA as opposed to individualism. In this regard, EAC and SADC should consider establishing a specialised institution mandated to coordinate the development of oil and gas infrastructure in the region similar to ECOWAS. Additionally, ECOWAS should make its specialised institution more inclusive by increasing the level of participation from its member states.

3. Strengthening the international legal framework to support regional cooperation

   The international community should escalate the involvement of regional organisations, specifically in addressing climate change and energy access. In this respect, the research in this book recommends that the Paris Agreement is amended to further strengthen and involve regional organisations in tackling both climate change and energy access in SSA.

Bibliography

Laws, Regulations and Treaties

Directive 2009/72/EC of the European Parliament and of the Council of 13 July 2009 concerning common rules for the internal market in electricity and repealing Directive 2003/54/EC (OJ L 211, 14.8.2009).
Directive 2009/73/EC of the European Parliament and of the Council of 13 July 2009 concerning common rules for the internal market in natural gas and repealing Directive 2003/55/EC (OJ L 211, 14.8.2009).
Electricity Act, Chapter 433 and the Petroleum Act, Chapter 434 of the Laws of Zambia.
Energy and Water Utilities Regulatory Authority (EWURA) Act Cap 414 of the laws of Tanzania.
Energy Regulatory Act, 2007 (Act No. 2 of 2007) Laws of Swaziland.
Energy Regulatory Act, 2011 (Chapter 13:23), Laws of Zimbabwe.
Energy Regulation Act, Cap 436 of the Laws of Zambia.
Energy Regulation Act 2004, Laws of Malawi.
Recital 33 of Directive 2009/72/EC and Recital 29 of Directive 2009/73/EC.
The Lesotho Electricity Act No. 12 of 2002 as amended, Laws of Lesotho.
The National Energy Regulator Act, 2004 (Act No. 40 of 2004), Laws of South Africa.

V. R. Nalule, *Energy Poverty and Access Challenges in Sub-Saharan Africa*, Energy, Climate and the Environment, https://doi.org/10.1007/978-3-319-95402-8

Books and Articles

Abeeku Brew-Hammond: Challenges to increasing Access to Modern Energy Services in Africa, codlogics_ff_201312 30_1388442782.report on West African energy ministers conference 2007.

Amin, M. (2008, July). Challenges in reliability, security, efficiency, and resilience of energy infrastructure: Toward smart self-healing electric power grid. In *Power and energy society general meeting-conversion and delivery of electrical energy in the 21st century, 2008 IEEE* (pp. 1–5). IEEE.

Armar, A. (2009). *Building regional power pools. A toolkit*. Washington, DC: The World Bank.

Auriol, E., & Biancini, S. (2015). Powering up developing countries through integration? *The World Bank Economic Review, 29*(1), 1–40.

Avila, N., Carvallo, J. P., Shaw, B., & Kammen, D. M. (2017). *The energy challenge in sub-Saharan Africa: A guide for advocates and policy makers*. Technical report, OXFAM.

Bazilian, M., Nussbaumer, P., Cabraal, A., Centurelli, R., Detchon, R., Gielen, D., et al. (2010). *Measuring energy access: Supporting a global target*. New York: Earth Institute, Columbia University.

Belguedj, M. (2006). *The West Africa gas pipeline and power pool project building blocks for regional energy integration*. A report prepared for the International Gas Union.

Bhatia, M., & Angelou, N. (2015). *Beyond connections energy access redefined*. Energy Sector Management Assistance Program, World Bank.

Board, M. A. (2007). *National security and the threat of climate change*. Alexandria: CNA Corporation.

Bonafé Martínez, E. (2010). *Towards a European energy policy: Resources and constraints in EU law* (Doctoral book).

Cameron, P. (2016). *Legal aspects of EU energy regulation: The consolidation of energy law across Europe*. Oxford University Press.

Cameron, P. D., & Brothwood, M. (2002). *Competition in energy markets: Law and regulation in the European Union*. Oxford University Press.

Cherp, A., & Jewell, J. (2014). The concept of energy security: Beyond the four As. *Energy Policy, 75*, 415–421.

Choi, J. B., Goo, B. K., Kim, J. C., Kim, Y. J., & Kim, W. S. (2003). Development of limit load solutions for corroded gas pipelines. *International Journal of Pressure Vessels and Piping, 80*(2), 121–128.

Choo, J. (2006). Energy cooperation problems in Northeast Asia: Unfolding the reality. *East Asia, 23*(3), 91–106.

Del Río, P., & Janeiro, L. (2016). Overcapacity as a barrier to renewable energy deployment: The Spanish case. *Journal of Energy, 2016.*

Dincer, I., & Rosen, M. A. (1998). A worldwide perspective on energy, environment and sustainable development. *International Journal of Energy Research, 22*(15), 1305–1321.

Diodorous Buberwa Kamala: *The achievements and challenges of the new East African Community Co-operation*. Retrieved from http://www2.hull.ac.uk/hubs/pdf/memorandum58.pdf

DoD, U. (2011). *Energy for the warfighter: Operational energy strategy*. Washington, DC: DD Deputy Secretary of Defense.

Downie, C., & Drahos, P. (2017). US institutional pathways to clean coal and shale gas: Lessons for China. *Climate Policy, 17*(2), 246–260.

Dupont, L. M., & Kuhlmann, H. (2017). Glacial-interglacial vegetation change in the Zambezi catchment. *Quaternary Science Reviews, 155*, 127–135.

EAPP Regional Power System Master Plan 2014: Volume II Data Report. Retrieved from http://www.ea-energianalyse.dk/reports/1332/1332_eapp_master_plan_2014_volume_2_data_report.pdf.

Eckholm, E. (1975). *The other energy crisis: Firewood*. Worldwatch Paper (1).

ECOWAS Centre for Renewable Energy and Energy Efficiency. Retrieved from http://www.ecreee.org/page/overview-ecreee.

ECOWAS Energy Policy A/DEC 3/5/82. Decision AIDEC 3/5/82 relating to the adoption of the ECOWAS Energy Policy, adopted by the Authority of Heads of State and Government in 1982.

Ellabban, O., Abu-Rub, H., & Blaabjerg, F. (2014). Renewable energy resources: Current status, future prospects and their enabling technology. *Renewable and Sustainable Energy Reviews, 39*, 748–764.

Ershad, A. M. (2017). Institutional and policy assessment of renewable energy sector in Afghanistan. *Journal of Renewable Energy, 2017,* 1–10.

Estache, A., & Wren-Lewis, L. (2009). Toward a theory of regulation for developing countries: Following jean-jacques laffont's lead. *Journal of Economic Literature, 47*(3), 729–770.

Ethier, W. (1998). *The international commercial system (No. 210)*. Princeton University International Economics.

EU-Africa Infrastructure Trust Fund, Annual Report 2010 (p. 26). Retrieved from http://www.eu-africa-infrastructuretf.net/attachments/Annual%20Reports/eu_africa_infrastructure_trust_fund_annual_report_2010_en.pdf.

FEWS NET: Southern Africa Special Report. 2014/15 EL Niño Event, July 2014.

Finer, M., Jenkins, C. N., Pimm, S. L., Keane, B., & Ross, C. (2008). Oil and gas projects in the western Amazon: Threats to wilderness, biodiversity, and indigenous peoples. *PLoS One, 3*(8), e2932.

G20 Energy Access Action Plan.

Gergacz, J. W. (1979). Solar energy law: Easements of access to sunlight. *New Mexico Law Review, 10*, 121.

Gippner, O. (2010). *Energy cooperation in South Asia: Prospects and challenges*. South Asia Watch on Trade, Economics & Environment (SAWTREE).

Gnansounou, E. (2008). Boosting the electricity sector in West Africa: An integrative vision. *International Association for Energy Economics, 17*, 23–29.

Goldthau, A. (2014). Rethinking the governance of energy infrastructure: Scale, decentralization and polycentrism. *Energy Research & Social Science, 1*, 134–140.

Hector G Sarmiento, Miguel Angel Avila: *Issues associated with international power grid interconnections in Mexico*.

Hettne, B., & Söderbaum, F. (2006). Regional cooperation: A tool for addressing regional and global challenges. In *Meeting global challenges: International cooperation in the national interest. Final report* (pp. 179–244).

Hettne, B., Sunkel, O., & Inotai, A. (Eds.). (2016). *Globalism and the new regionalism* (Vol. 1). Springer.

Hilton, I., & Kerr, O. (2017). The Paris Agreement: China's 'New Normal' role in international climate negotiations. *Climate Policy, 17*(1), 48–58.

Huber, M. (2017). *Carbon nation: Fossil fuels in the making of American culture*. By Bob Johnson.

Huda, M. S., & McDonald, M. (2016). Regional cooperation on energy in South Asia: Unraveling the political challenges in implementing transnational pipelines and electricity grids. *Energy Policy, 98*, 73–83.

International Energy Agency: World Energy Outlook, 2014.

International Energy Agency: World Energy Outlook Special Report, 2014.

International Energy Agency: World Energy Outlook-Methodology for Energy Access Analysis. Retrieved from www.worldenergyoutlook.org/media/weowebsite/EnergyAccess_Methodology_2014.pdf.

Kanagawa, M., & Nakata, T. (2006). Analysis of the impact of electricity grid interconnection between Korea and Japan—Feasibility study for energy network in Northeast Asia. *Energy Policy, 34*(9), 1015–1025.

Karekezi, S., Kimani, J., Mutiga, A., & Amenya, S. (2003). *Energy services for the poor in Eastern Africa: Sub-regional "Energy Access" study of East Africa*. A report prepared for "Energy Access" working group of the Global Network on Energy for Sustainable Development.

Keating, M. (1998). *The new regionalism in Western Europe: Territorial restructuring and political change*. Cheltenham, UK: E. Elgar.

Kennedy, J. L. (1993). *Oil and gas pipeline fundamentals*. Pennwell Books.

King, M. D., & Gulledge, J. (2014). Climate change and energy security: An analysis of policy research. *Climatic Change, 123*(1), 57–68.

Kriegler, E., Bauer, N., Popp, A., Humpenöder, F., Leimbach, M., Strefler, J., et al. (2017). Fossil-fueled development (SSP5): An energy and resource intensive scenario for the 21st century. *Global Environmental Change, 42*, 297–315.

Lazkano, I., Nøstbakken, L., & Pelli, M. (2017). From fossil fuels to renewables: The role of electricity storage. *European Economic Review, 99*, 113–129.

Lee, C. K. F., Duncan, C., Owen, H. J. F., & Pettorelli, N. (2018). A new framework to assess relative ecosystem vulnerability to climate change. *Conservation Letters, 11*(2), e12372.

Lovering, J. (1999). Theory led by policy: The inadequacies of the 'new regionalism' (illustrated from the case of Wales). *International Journal of Urban and Regional Research, 23*(2), 379–395.

Lubell, H. (1961). Security of supply and energy policy in Western Europe. *World Politics, 13*(3), 400–422.

Mathews, J. A., & Tan, H. (2014). Manufacture renewables to build energy security. *Nature, 513*(7517), 166.

McNutt, M. (2013). Climate change impacts. *Science (New York, NY), 341* (6145), 435.

Mwakyusa, A. (2016, March 8). East Africa: Construction of Uganda—Tanga port oil pipeline starts August. *Tanzania Daily News via AllAfrica.com*. Dar es Salaam. Retrieved March 8, 2016.

NASA. Global Climate Change. Retrieved from https://climate.nasa.gov/effects/.

Nepal, R., & Jamasb, T. (2012). Interconnections and market integration in the Irish Single Electricity Market. *Energy Policy, 51*, 425–434.

Niang, I., Ruppel, O. C., Abdrabo, M. A., Essel, A., Lennard, C., Padgham, J., et al. (2014). Africa. In *Climate change 2014: Impacts, adaptation, and vulnerability. Part B: Regional aspects*. Contribution of Working Group II to the Fifth Assessment Report of the Intergovernmental Panel on Climate Change [Barros, V.R., C.B. Field, D.J. Dokken, M.D. Mastrandrea, K.J. Mach, T.E. Bilir, M. Chatterjee, K.L. Ebi, Y.O. Estrada, R.C. Genova, B. Girma, E.S. Kissel, A.N. Levy, S. MacCracken, P.R. Mastrandrea, and L.L. White (eds.)]. Cambridge University Press, Cambridge, United Kingdom and New York, NY, USA, pp. 1199–1265. Can be retrieved from http://www.ipcc.ch/pdf/assessment-report/ar5/wg2/WGIIAR5-Chap22_FINAL.pdf.

Nizami, A. S., Shahzad, K., Rehan, M., Ouda, O. K. M., Khan, M. Z., Ismail, I. M. I., et al. (2017). Developing waste biorefinery in Makkah: A way forward to convert urban waste into renewable energy. *Applied Energy, 186*, 189–196.

Nkhonjera, G. K. (2017). Understanding the impact of climate change on the dwindling water resources of South Africa, focusing mainly on Olifants River basin: A review. *Environmental Science & Policy, 71*, 19–29.

Ogwezzy, M. C. (2017). Legal framework for the implementation of ECOWAS energy programmes: The energy protocol, WAPP Agreement and WAGP Treaty. *Oil, Gas & Energy Law Journal (OGEL), 15*(1).

Painuly, J. P. (2001). Barriers to renewable energy penetration; a framework for analysis. *Renewable Energy, 24*(1), 73–89.

Parail, V. (2010). Properties of electricity prices and the drivers of interconnector revenue. *Cambridge Working Paper in Economics 1059/Electricity Policy Research Group 1033*. University of Cambridge, UK.

Prosser, T. (1997). *Law and the regulators*. Oxford University Press.

Quinn, E. L. (2009). Privacy and the new energy infrastructure. Retrieved from https://papers.ssrn.com/sol3/papers.cfm?abstract_id=1370731.

REN21, Renewables 2014: Global Status Report, REN21 Bonn, Germany 2014.

SADC, Energy Monitor, 2016.

SADC Renewable Energy and Energy Efficiency Strategy and Action Plan (REESAP 2016–2030).

Schiff, M., & Winters, L. A. (2002). Regional cooperation, and the role of international organisations and regional integration. *World Bank Policy Research Working Paper (2872)*.

Serur, A. B., & Sarma, A. K. (2017). Impact of spatial data availability on climate change prediction in the Weyib River Basin in Ethiopia. *Water Resources Management, 31*(6), 1809–1824.

Shahriar, A., Sadiq, R., & Tesfamariam, S. (2012). Risk analysis for oil & gas pipelines: A sustainability assessment approach using fuzzy based bow-tie analysis. *Journal of Loss Prevention in the Process Industries, 25*(3), 505–523.

Spalding-Fecher, R., Joyce, B., & Winkler, H. (2017). Climate change and hydropower in the Southern African Power Pool and Zambezi River Basin: System-wide impacts and policy implications. *Energy Policy, 103*, 84–97.

Stiles, G., & Murove, C. (2015). *SADC renewable energy and energy efficiency status report*, Page 9.

Tabatabaei, M., Karimi, K., Kumar, R., & Horváth, I. S. (2015). Renewable energy and alternative fuel technologies. *BioMed Research International, 2015*.

The Secretary General's Advisory Group on Energy and Climate Change: Energy for a Sustainable Future, Report and Recommendations, 28th April 2010, New York. Retrieved from http://www.un.org/millenniumgoals/pdf/AGECCsummaryreport[1].pdf.

The World Bank: Fact Sheet: The World Bank and Energy in Africa. Retrieved from http://web.worldbank.org/WBSITE/EXTERNAL/COUNTRIES/AFRICAEXT/0,,contentMDK:21935594~pagePK:146736~piPK:146830~theSitePK:258644,00.html.

Uddin, N., & Taplin, R. (2015). Regional cooperation in widening energy access and also mitigating climate change: Current programs and future potential. *Global Environmental Change, 35*, 497–504.

UN General Assembly Resolution 65/151, International Year of Sustainable Energy for All. Retrieved from http://www.un.org/ga/search/view_doc.asp?symbol=A/RES/65/151. The Resolution was passed on 20th December 2010.

UN Secretary-General Ban Ki-moona, Sustainable Energy for All. Retrieved from http://www.se4all.org/our-vision/.

UN-Energy: The energy challenge of achieving the millennium development goals. Retrieved from http://www.un-energy.org/publications/50-the-energy-challenge-for-achieving-the-millennium-development-goals.

United Nations Foundation: Achieving Universal Energy Access. Retrieved from http://www.unfoundation.org/what-we-do/issues/energy-and-climate/clean-energy-development.html.

United Nations: *Multi dimensional issues in international electric power grid interconnection*, 2011.

USAID, Southern Africa Drought, Fact Sheet, Jan 2017. It can be retrieved from https://scms.usaid.gov/sites/default/files/documents/1866/southern_africa_dr_fs04_01-30-2017.pdf.

Valeri, L. M. (2009). Welfare and competition effects of electricity interconnection between Ireland and Great Britain. *Energy Policy, 37*(11), 4679–4688.

Yeeles, A., & Akporiaye, A. (2016). Risk and resilience in the Nigerian oil sector: The economic effects of pipeline sabotage and theft. *Energy Policy, 88*, 187–196.

Yergin, D. (2006). Ensuring energy security. *Foreign Affairs, 85*(2), 69–82.

Index

V. R. Nalule, *Energy Poverty and Access Challenges in Sub-Saharan Africa*, Energy, Climate and the Environment, https://doi.org/10.1007/978-3-319-95402-8

www.ingramcontent.com/pod-product-compliance
Ingram Content Group UK Ltd.
Pitfield, Milton Keynes, MK11 3LW, UK
UKHW020956050925
462611UK00012B/1017
* 9 7 8 3 0 3 0 0 7 0 2 9 8 *